Wrecked on the Reef

WRECKED *on the* REEF

Maritime Archaeology of American Whaleships in the Pacific Ocean

JASON T. RAUPP

The University of Alabama Press
Tuscaloosa

The University of Alabama Press
Tuscaloosa, Alabama 35487-0380
uapress.ua.edu

Typeface: Adobe Caslon Pro

Book design: Sandy Turner Jr.

Cover image: *A Shoal of Sperm Whale, off the Island of Hawaii*, lithograph of painting by Thomas Birch; courtesy of the New Bedford Whaling Museum
Cover design: Terrell Harris

Cataloging-in-Publication data is available from the Library of Congress.
ISBN: 978-0-8173-2251-9 (cloth)
ISBN: 978-0-8173-6231-7 (paper)
E-ISBN: 978-0-8173-9580-3

To Jen McKinnon and Abe Raupp, whose love, support, and extraordinary patience made this book possible

Contents

Illustrations

Figures

Tables

Acknowledgments

The research and publication of this book were made possible by the support and encouragement of numerous individuals and organizations. Although the names of the skilled librarians, archivists, museum staff, academics, and heritage professionals are too numerous to list here, their assistance and impact on this work cannot be understated.

Much of the data used for this book resulted from research conducted at Flinders University (South Australia). I would like to express gratitude to the many other members of the Flinders University Department of Archaeology faculty, staff, and friends for their assistance, and to Professors Wendy Van Duivenvoorde and Heather Burke for their enthusiastic guidance and invaluable input. I am also grateful for research funding offered by the Flinders University Office of Graduate Research, the Australian Postgraduate Award scholarship program, and the Faculty of Humanities, Education and Law.

The management and staff at Papahanaumokuakea Marine National Monument and the NOAA National Marine Sanctuaries program provided generous support and access. In particular, I would like to express thanks to then Maritime Heritage coordinator Kelly Keogh for her friendship, for her steadfast support of this project, and for providing me numerous opportunities to assist in the exploration of the incredible maritime heritage of the Northwestern Hawaiian Islands. Numerous other colleagues contributed to the scientific and archival research of the whaling shipwreck sites in the monument. And finally, thanks to the officers and crew of the now decommissioned NOAA ship *Hi'ialakai* for maintaining a safe, sound, and welcoming environment onboard.

Collections housed in numerous libraries and archives in Australia and the United States were consulted for this research. Among the many facilities visited were the National Library of Australia, State Library of South Australia, Flinders University Library, State Library of Victoria, Australian National University libraries and Pacific Manuscripts Bureau, Vaughn Evans research library of the Australian National Maritime Museum, Massachusetts

Historical Society, Providence Public Library, Hawaii State Archives, Hawaiian Mission Houses Museum and Archives, and University of Hawaii's Hamilton Library and Pacific Collection. Of particular note are the internationally renowned repositories that focus on whaling, including New Bedford Whaling Museum Research Library, New Bedford Free Public Library, Nantucket Historical Association Research Library, Martha's Vineyard Museum, and Mystic Seaport.

Several colleagues provided input or support to this project over the course of this research, and to each of them, I owe a debt of gratitude. These individuals include the late Terry Arnott of South Australia's Department for Environment and Heritage; Peter Harvey, Cassandra Philippou, Hannah Steyne, and others at Heritage Victoria; independent researcher Adam Woolfe for sharing information about the whaling heritage of Western Australia, as well as his helpful tips for conducting research in New Bedford and Nantucket; Associate Professor Martin Gibbs of the University of Sydney for advice on conducting whaling research in Australia; Adam Patterson of Flinders University for the many discussions of historic whaling technologies; and to all my friends at the Western Australian Maritime Museum for providing support, advice, and information that proved useful to this research. And though not directly related to whaling, I would like to express my great appreciation for the generosity and advice provided by Emeritus Professors Richard Gould and Patrick Malone of Brown University.

And finally, a heartfelt thanks goes to my family and friends for their unwavering support and encouragement.

Wrecked on the Reef

Introduction

Organizing a colonial whaling enterprise was a priority of early English settlers to North America. The frequency of coastal whale stranding meant that colonists could easily undertake opportunistic whaling, which involved simply processing carcasses and storing whale products for shipment to markets. The profitability of these early activities and the knowledge that North Atlantic right whales (*Eubalaena glacialis*) were found easily just offshore led to the establishment of shore whaling enterprises (Allen 1928:340). The success of Atlantic shore whaling in North America soon brought an increased demand for whale products, and by the early eighteenth century, pelagic whaling methods developed. Pelagic whaling involved outfitting sailing vessels with small boats and sending them out in search of prey; if successful in the hunt, whale carcasses were towed back to a "mother ship," where blubber was removed and stored for later processing on shore. Despite the success of pelagic whaling, the small cargo capacities of the vessels employed as mother ships restricted the number of whales that could be captured on a voyage. Furthermore, warm temperatures tended to spoil unprocessed blubber, so whaling cruises could only be conducted as seasonal operations or in consistently cold climates. A shift to hunting higher-valued sperm whales (*Physeter macrocephalus*) and the growing need to make longer voyages to more distant hunting grounds led to the transfer of blubber processing onto to the decks of whaleships around 1750 (Davis et al. 1997:36). Though seemingly simple in concept, this American innovation sparked a revolution in whaling and produced a multinational industry that was truly global in scope.

By the early nineteenth century, American whaleships entered the Pacific Ocean and began exploring its vast expanse. In many areas, they found and exploited immense numbers of sperm whales. The location of new hunting grounds, and an understanding of the seasonal movements of different whale species, brought increasing numbers of ships that roamed throughout the region year-round. European markets found the massive quantities of extracted whale oil the perfect fuel for the industrial revolution and used it to lubricate machines and illuminate factories. These applications allowed for increased productivity and in turn created greater demand for whale products. As a result, the whaling firms of New England soon sent most of their fleets on extended voyages to remote and often uncharted areas of the Pacific Ocean in the hope of full cargo holds and quick returns.

This book explores the industrial nature of America's pelagic whaling ships operating in the Pacific Ocean from the early to mid-nineteenth century. Unlike the vessels employed in other seaborne merchant activities of the period, whaleships were constructed and organized to be floating production platforms. The industrial system employed involved organized labor and a series of interrelated processes and tools, many of which were adapted from European whaling traditions that developed over centuries. By connecting historical and archival research, archaeological site inspections, and comparative studies of museum collections, this book contextualizes the industrial experience and working environment that existed onboard those whaling vessels. Each of these sources of information provides a different perspective, as seen through the recognition of the material components of the fishery and the ways in which technologies were employed. The integration of these datasets helps to illuminate many of the social factors that influenced these highly successful industrial workplaces of the "golden age of whaling" (1815–1860).

Whaling as a Standardized Industrial System

Pelagic whaling practices changed little from the time of their establishment in the mid-eighteenth century until new technologies were introduced beginning in the mid-nineteenth century. Wooden sailing ships cruised known grounds looking for prey; when sighted, small whaleboats were lowered to pursue them and quickly maneuvered close enough to attack. A harpoon attached to a long line was thrown by hand and stuck into the whale, which then dragged the boat and crew until it tired. At that point, the whale was killed using a sharp lance, and its carcass was towed back to the waiting ship where it was processed (Whitehead 2003:14). Processing involved several stages, including removing the blubber and case oil, rendering the blubber into oil, transferring oil into casks for storage and transport, and cleaning the work areas to prepare for the next round of activity.

The ships used for pelagic whaling during this period were workplaces that incorporated complex industrial processes that resulted from wider social, cultural, and technological changes throughout the early to mid-nineteenth century. Unlike vessels employed in other maritime trades, whaling ships were self-contained and fully integrated working platforms, as well as complex social systems. They contained both the equipment necessary to carry out whaling operations and the domestic spaces that provided a meager home for officers and crews for multiple years. Equipped for long voyages, whaleships had no need to touch land for any other purpose than resupplying when fresh provisions were depleted or emptying holds of oil for transshipment when an opportunity presented itself.

The success of their endeavors depended not only on standardized processes and technologies but also on the effective management of organized labor. While searching for prey, order was maintained through an established system of watches. During these shifts, crews kept busy with tasks associated with ship keeping, gear maintenance, and lookout duties. Officers of the watch supervised these activities and assisted captains in making decisions about best plans of action. When whales were spotted, a flurry of activities erupted onboard as boats were lowered for pursuit and the ship was readied for blubber processing. Unlike other fisheries that depended on getting fresh, frozen, or salted fish to market in a relatively short amount of time, whalers signed on to voyages that lasted until the ship's hold was full or its captain decided to return to port. To ensure ultimate productivity, American whaleship owners relied on a wage system that paid each crew member a prearranged fractional share of the total net proceeds of a voyage (Hohman 1926:644).

Pelagic whaling ships of the nineteenth century supported the industrial processes that operated in conjunction with one another for commercial success and survival. Further, these ships provided structure for a community of crew members who lived and worked together for extended periods of time. They relied on each other for the success of their catch, as well as the overall success of the cruise and their share of the profits; but in many instances, they also relied on each other for their lives. As the hunting grounds grew more distant and voyages became longer, standardization of industrial practices and techniques onboard pelagic whaleships became necessary for success. This standardization also manifested in changes to the overall design and layout of the whaling vessel and its component parts so that the processes involved could be undertaken with greater efficiency. By the early nineteenth century, the basic pattern of many of these technological features would have distinguished ships as whale hunters, irrespective of the nationality of the vessel.

Although some observers of the industry refer to whaleships of the early nineteenth century as "floating oil factories," by the very nature of the activities supported onboard, the concept of a factory is unsuitable (Allen 1973:159; Day 1966:132; Weiss et al. 1974:79). The term *factory* is a contraction of the word *manufactory* and is a physical structure where goods are commonly manufactured by machine (Casella 2001:25). When applied to whaleships, this term then implies that onboard industrial activities involved secondary manufacturing processes. Pelagic whaling, however, was an extractive industry, and the operations that took place onboard were primary in nature. Much like in other extractive industries, such as mining or forestry, early nineteenth-century whaleships were responsible for securing raw materials, and the processing that occurred onboard the vessels simply converted them into a form that allowed for maximum storage capacity and ease of transport. The products procured from sperm and other whale species included varying grades of oil, occasional ambergris, and, to a lesser extent, bone and teeth. Each of those materials required some degree of secondary processing and/or refinement before they were made commercially available.

Except for the ship's windlass, which was primarily a component of its anchoring system that was adapted for use in blubber processing, pelagic whaling was unmechanized during the early nineteenth century (Gordon and Malone 1994:228). The onboard industrial processes changed little from the time of their introduction in the eighteenth century and only targeted specific parts of the whale. From the 1860s, however, the whaling industry experienced a series of technological enhancements that resulted in increased mechanization. Advances in harpooning technologies and the introduction of steam power to some vessels allowed whalers to hunt species that were previously considered too fast for conventional whaling methods (Clapham and Baker 2002:1328). Such changes became a catalyst for the development of the modern whaling era and would ultimately lead to the development of floating factory ships in the twentieth century (Tønnessen and Johnsen 1982).

As a combined tanker and factory, the large steel-hulled factory ships employed in modern whaling are dramatically different from their wooden predecessors of the mid-nineteenth century. Fully mechanized, they included all equipment necessary for efficiently processing entire whale carcasses into final products ready for commercial sale (Basberg 1998). Similar to their predecessors, they acted as mother ships for a small fleet of steam-powered whale catchers that explored surrounding seas in search of whales. Once spotted, the whale catchers used explosive rocket harpoons mounted on their bows to kill whales and then tow carcasses back to the factory ships for processing. Thus, the term *factory* is more fitting of the vessels employed in twentieth-century pelagic whaling, as they included manufacturing capabilities.

Pelagic Whaling as Industrial Archaeology

To understand the systems that operated onboard pelagic whaleships of this period, relevant data is viewed through the lens of industrial archaeological practice and theory. Industrial archaeology is an academic subdiscipline that has grown into a vibrant and progressive area of research and practice over the past six decades (Basberg 2004:22; Cossons 2005:ix). Although early industrial archaeological studies focused primarily on recording and preserving the technological and structural remains of the industrial revolution, as the subdiscipline matured, more consideration was given to how social and cultural factors might be incorporated into its agenda. The primary aims of more recent archaeological investigation into sites of past industrial activity include identifying not only technical processes but also human activities that took place, when and where on sites they occurred, and how they affected and were affected by technical processes (Badcock and Malaws 2004; Cranstone 2001; English Heritage 2006). This type of inquiry provides a unique perspective on pelagic whaling because it focuses on the ship as an industrial platform; on the whalers and their day-to-day lives onboard the vessel; and on the setting, or seascape, in which this activity occurred. All too often past maritime archaeological studies have focused solely on aspects of ships, such as construction techniques or the technology employed to operate them (Adams 2001; Flatman 2003). The application of an industrial archaeological approach to pelagic whaling, however, expands beyond the ship to examine pelagic whaling as an integrated system inclusive of the human experience.

Though prominent industrial archaeologists in the relatively recent past (see Gordon and Malone [1994]) suggested the validity of considering early nineteenth-century pelagic whaleships as industrial archaeological sites, to date no such analysis is known. Perhaps the most closely related was Bjørn Basberg's (1998) examination of twentieth-century whaling factory ships. That work only briefly considers eighteenth- and early nineteenth-century whaleships as predecessors and instead focuses on the technological developments of the latter part of the nineteenth century that produced the modern whaling industry (Basberg 1998:23). Basberg was especially interested in the developmental phases and specific features of twentieth-century Antarctic factory ships that led to the establishment of a "dominant design" (Abernathy and Utterback 1978; Basberg 1998:22). While his work presents a thorough analysis of industrial development and ship design for that era, Basberg also inadvertently highlights the need to research the evolution of earlier whaleships, their technology, and industrial relations to better understand the historical development of the pelagic whaling industry as a whole.

This book therefore investigates early pelagic whaleships as industrial workplaces and their evolution in design and technology through historical and archaeological inquiry by focusing on pelagic whaling as a resource extraction industry. Further, it explores the industrial seascape within which this fishery operated, as well as how the industry and its actors shaped and affected it. To better understand these workplaces, the text explores historical documentation and material remains of whaleships wrecked in the early nineteenth century in relation to available technologies of the time and specific processes of hunting, capturing, processing, storing, and transporting whale products. Each stage in the whaling process is represented by a distinctive toolkit and set of procedures that developed along a trajectory to produce the fully integrated maritime industry of shipboard whaling conducted in the early nineteenth century. That, in turn, culminated in the extractive process occurring solely onboard the ship and away from shore.

Pelagic whaleships were floating work platforms that functioned under an organized framework and utilized many tools specific to that trade. Therefore, the well-preserved remains of wrecked vessels present an opportunity to gain insight into onboard industrial operations. Additionally, much of the maritime archaeological research from the past 50-plus years relied heavily on particularist approaches to wreck site investigations, focusing mainly on technological aspects of ships. While such studies resulted in much valuable information, too often they arrived at their conclusions by bypassing the human element that shaped those remains (Adams 2001; Dellino-Musgrave 2006; Flatman 2003; Hocker 2004). Thus, bridging the subdisciplines of maritime and industrial archaeology and applying an approach that is inclusive of social and cultural aspects reveals information about the daily life of the communities that existed onboard whaleships and the technologies employed.

Book Organization

Chapter 1 explores interpretive methods used by industrial archaeological researchers and their conceptual adaptation to ships employed in the pelagic whale fishery. An overview of previous archaeological investigations of wrecked whaleships illustrates the paucity of scholarly studies focused on this aspect of the trade. To better understand pelagic whaling as an extractive activity, interpretive themes derived from industrial archaeological studies are considered. The conceptual adaptations include *maritime industrial workplaces*, *marine resource extraction*, and *industrial seascapes* and are defined to understand the ways in which the wrecks of nineteenth-century pelagic whaling vessels benefit from interpretation as industrial sites.

Chapter 2 provides a general overview of the history of whaling from its beginnings as an opportunistic coastal pursuit to the golden age of commercial whaling in the early to mid-nineteenth century. As whaling was an important component in the development of modern American society, the subject has long been a focus for historians and economists. From as early as 1820, authors have dedicated volumes to understanding the intricacies of the whale fishery. While all aspects of the trade are undoubtedly important, pelagic whalers operated in a specific environmental setting. Thus, this chapter does not purport to be a definitive history of whaling; instead, it provides an overview of American whaling with specific emphasis on activities in the Pacific Ocean during the nineteenth century.

Chapter 3 offers a discussion of the development of Hawaii as an entrepôt for the whaling trade in the central Pacific Ocean. Upon crossing from the Atlantic to the Pacific via Cape Horn (now modern Chile) in the late eighteenth century, whalers found immense cetacean populations. The ensuing rush to harvest those resources caused a rapid depletion of stock, which led to exploration of increasingly distant areas. A pattern of discovery and depletion followed, resulting in voyages of ever-increasing length. To accommodate the extended time at sea, agents developed a system of outfitting that provided whaleships with every conceivable item that might need to be replaced. The need for fresh food and water, however, was a consistent concern and led captains to chart safe islands for reprovisioning when encountered. By 1820, Hawaii emerged as an ideal island group due to its geographic location. Since the Hawaiian archipelago was not without danger, however, many whaling vessels came to grief on its barely submerged reefs.

Given the vast expanse of the Pacific Ocean, it is no surprise that numerous nineteenth-century whaleships wrecked on or around its many reefs and atolls. Factors ranging from uncharted physical dangers to storms to navigational errors all contributed to losses and left vessel remains scattered over wide areas. Though such remains hold tantalizing evidence of these once floating maritime industrial sites, their generally remote locations keep them out of reach from archaeologists. Exceptions to this, however, are the wrecks situated with the boundaries of Papahanaumokuakea Marine National Monument (PMNM) in the Northwestern Hawaiian Islands (NWHI) (Figure I.1). As such, chapter 4 presents the archaeology of American whaleships wrecked within that region, as well as an analysis and interpretation of their material remains.

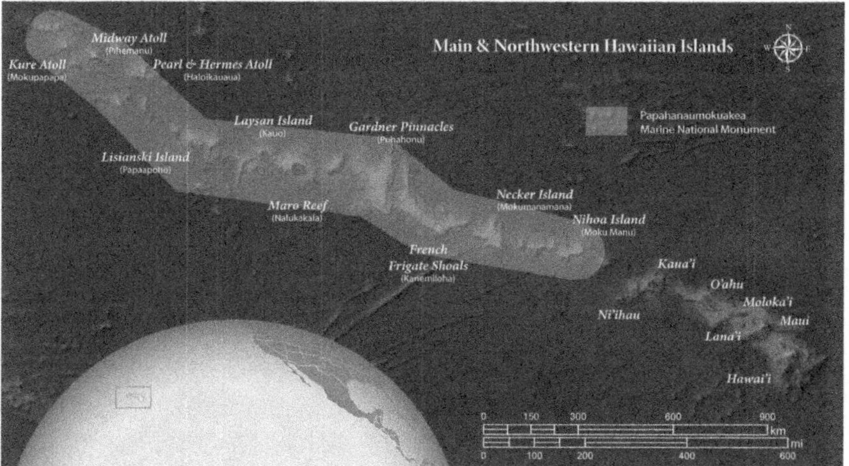

Figure I.1. Northwestern Hawaiian Islands run in an arc to the northwest of the main Hawaiian archipelago and are encompassed with the boundaries of Papahanaumokuakea Marine National Monument. (Courtesy of Jeremy Borrelli.)

Early nineteenth-century pelagic whaling developed from mid-eighteenth-century industrial organization and technological innovation. Along with the tools and techniques employed by whalers of the period came a regimented and sequential system for operations. Incorporating a hierarchical labor model that included defined roles, an incentivized wage structure, and risk management, the American system eventually dominated the fishery. Chapter 5 describes the system under which pelagic whaling voyages operated and focuses specifically on how preparations made prior to departure, as well as those undertaken upon arrival in home ports, helped to produce the distinct archaeological signatures of the whaleship losses investigated in the NWHI.

Expanding on the discussion of the American whaling system, chapter 6 presents a historical and archaeological discussion of the different aspects of whaling cruise operations. Though no single phase of the system was more important than the others, it was during this middle period, when whale hunting and processing occurred, that many of the activities described in popular literature on the subject took place. To better understand the daily routines that played out on these ships, topics such as work and recreation periods, food, crew health, and industrial methods are all explored in relation to the many whaling-specific artifacts identified at relevant shipwreck sites.

The conclusion ties together all the data presented and explores the interpretation of whaling shipwrecks as maritime industrial sites. Unlike most commercial ships of the period, whaling vessels carried the equipment and crew needed to extract a specific resource from the sea at an industrial scale. As floating industrial sites, the remains of such vessels can provide perspective on themes including organizational structure, onboard conditions, hunting methods and equipment, operational scale, and environments encountered. Through the adaptation of interpretative methods from the subdiscipline of industrial archaeology, new perspectives for wrecked whaling ships emerge.

1

Archaeological Investigations of Whaleships as Workplaces

The early nineteenth century was a dynamic period in which standardization of technology and organization of labor through defined roles resulted in efficient industrial operations. The processes involved allowed for shipboard whale oil production to take place on an industrial scale in an otherwise uncertain and often uncharted seascape. The ships engaged in early nineteenth-century pelagic whaling were microcosms of industrial life, but much like frontier mining sites of the American West, each represented only a piece of the larger capitalistic pursuit of an industrial whole. Thus, the exploration of industrialization in pelagic whaling offers insight into the social, cultural, and technical conditions that existed onboard the vessels employed during the industry's golden age (Davis et al. 1988:570; Davis et al. 1997:38; Hegarty 1964; Tompkins 1972:77). This chapter explores the conceptual adaptation of interpretive themes developed by practitioners of industrial archaeology to understand whaleships as industrial sites.

Previous Archaeological Investigations of Whaleships

The investigation of wrecked whaleships provides insightful information about the types of vessels used and the different technologies employed for nautical whaling operations. Ranging from the sixteenth century to the late nineteenth century, the remains of ships associated with the industry have been identified in several areas of the world and exist in various states of preservation. The whaleship, or *whaler* as it is commonly referred, is a type of craft that operated as a mother ship by supporting small whale-hunting boats, transporting the catch to shore or the ship for processing, and storing oil for eventual shipment to markets. Although many other vessel types were involved in whaling over the centuries, those are not the focus of this book. They are, nevertheless, relevant from a comparative standpoint and are briefly discussed here to illustrate the continuum of whaling and whaleships as a theme within maritime archaeology.

By far the most extensively investigated wreck of a whaling vessel is that of the sixteenth-century galleon found in Labrador, Canada. This shipwreck site is located just offshore from the remains of Basque shore-whaling operations in Red Bay (Logan and Tuck 1990). Using documentary sources as a guide, Parks Canada archaeologists discovered the remains of the ship in 1978. Over the course of the next six years, two additional Basque whaleships and four whaleboats—known as *chalupas*—were found in the clear, cold waters of Red Bay (Grenier et al. 2007; Stevens 1997:336). It was the first shipwreck identified, however, that received the most attention. The wreck is most likely that of the whaling galleon *San Juan*, which sank in the autumn of 1565 with nearly 2,000 barrels of whale oil stowed onboard (Barkham 1984:516). The remains of the vessel were studied extensively over several field seasons and systematically excavated, mapped, dismantled, brought to the surface for precise recording, and then reburied on site (Grenier 1988:70; Grenier et al. 2007). Many aspects of sixteenth-century nautical activities were documented, including navigation, ship construction, iron working, fittings, and rigging, as well as whaling technologies used by the Basque in Labrador, and life onboard their ships (Barkham 1981; Bradley 1993; Grenier 1988; Grenier et al. 2007; Light 1990, 1992; Ross 1985; Wadell 1985, 1986).

Over the past 45 years, archaeologists have also documented the wrecks of whaling ships in both the Indian and Pacific Oceans. Of those sites, the remains of several pelagic whaleships have been identified and studied along the coast of western Australia. The wrecks include those of the British whaler *Lively*, lost on Rowley Shoals in 1811 (Atkinson 1987; Henderson 1983; Nutley 1987; Stanbury 2015); the French-colonial whaling barque *Lancier*, wrecked off Stragglers Reef in 1839 (Kenderdine 1994); the American whalers *North America*, *Samuel Wright*, and a second *North America*, all blown ashore at Bunbury between 1840 and 1843 (Anderson and McAllister 2012; McAllister 2012; McCarthy 1982); the American whaleship *Cervantes*, lost at Jurien Bay in 1844 (McAllister 2013); the Australian whaling barque *Lady Lyttleton*, wrecked near Albany in 1867 (Vosmer and Wright 1991); and the American whaler *Day Dawn*, lost near Perth in 1886 (Kimpton and Henderson 1991; McCarthy 1981). Staff and volunteers of the Department of Maritime Archaeology at the Western Australian Museum recorded each of those sites, and, in most cases, a small number of artifacts were collected and curated.

Although government mandate and management considerations provided the impetus for most of the investigations, academic pursuits prompted specific research at some sites. The wrecks of *Day Dawn* and *Lady Lyttleton* were the focus of a postgraduate maritime archaeological field course and resulted in detailed analyses of artifact assemblages (Erskine 1997a, 1997b), ship construction (Vosmer and Wright 1991), site deterioration (Thomson 1997), and cultural heritage management research (Moran 1997; Williams 1997). Furthermore, investigations of the wrecks at Bunbury in the southwest of the state supplied data for a master's thesis that proposed a typology for nineteenth-century whaleships (McAllister 2012).

Two other shipwrecks documented along the Australian coastline are thought to have been involved in pelagic or bay whaling at some point in their careers. These are the Hobart whaler *Litherland*, lost in Bass Strait off Tasmania in 1853 (Nash 1990), and the barque *Cheviot*, wrecked off Wilson's Promontory, Victoria, in 1854 (Anderson 2004). Though both vessels were equipped for whaling (e.g., the remains of tryworks and other industrial tools were documented), historical data indicates that neither was actively involved in the trade at the time of wrecking.

Although the largest concentration of pelagic whaleship wreck sites was documented around Australia, several others have been investigated in scattered locations throughout the Pacific. The wrecks of two whaleships were identified in the Federated States of Micronesia (FSM) state of Kosrae. One of these—possibly the remains of the British whaler *Harriet* lost in 1843 (Ward 1960:525)—was recorded by the US National Park Service (Silva et al. 1992). The remains of three other whaleships were examined and preliminarily documented in Madolenihmw Harbor of the FSM state of Pohnpei. Though only preliminarily investigated, they are thought to represent three of four American whaleships attacked and sunk by the Confederate raider *Shenandoah* in 1865 (Finney and Graves 2002).

Technically not shipwreck sites, the remains of three other pelagic whaleships in the Pacific have been archaeologically investigated. Each of these was fitted out and employed as a whaler early in its career but later reused for other purposes. Two of them—the former whaleships *Niantic*, built in Chatham, Connecticut, in 1835, and *Candace*, built in 1818 in Boston, Massachusetts—were sold to interests in San Francisco, California, and converted to storeships on the city's waterfront (Delgado 1980; Strother et al. 2007). The extremely well-preserved lower hull sections of each of those vessels were exposed and thoroughly documented as the result of modern development activities (Delgado 2006, 2009; Strother et al. 2007). The American whaleship *Othello* was another whaling vessel removed from service and repurposed. Built in 1853, *Othello* sailed for a decade as a whaler before being sold first to interests in Sydney, Australia, and then to a shipyard on Stewart Island, New Zealand, to be scuttled at the outer end of a T-wharf. The well-preserved remains of the ship's lower hull lie in situ at the former shipyard and were documented in 2012 (Dodd 2013).

Whaleships and Maritime Archaeology

Due to both their nautical nature and their depositional environments, investigation of ship remains typically falls under the purview of maritime archaeology. Defined by Keith Muckelroy (1978:4) as "the scientific study of the material remains of man and his activities on the sea," maritime archaeology incorporates the principals and practices used to investigate terrestrial archaeological sites and applies them in submerged environments.

Of course, in elementary form, whalers are like other wooden sailing vessels of the period—solidly built wooden hulls, sturdy masts and spars, extensive standing and running rigging, and expansive sail patterns all provided their basic features. The difference between wrecked whaleships and other vessels lies in the interpretation of their function and purpose. Either converted from other industries or purpose-built, the hulls of early nineteenth-century whaleships housed specific technological features that clearly distinguished them from other ships and performed the function of catching and processing whales. As such, these vessels present a unique opportunity to investigate a specific maritime industrial process and workplace.

Much like other subdisciplines within archaeology, maritime archaeology has experienced growing pains associated with finding its place within the broader discipline. Developing out of historical particularist approaches to the remains of ships, early maritime archaeological projects focused primarily on the potential for shipwrecks to expose data pertaining to patterns in ancient seafaring (Bass 1983). In the 1960s, expensive excavations of a few sites were undertaken in which wrecks were recorded in minute detail on the seabed before all or parts of them were recovered for further analysis in laboratories on land (Bass 1967; Cederlund 2006; Crumlin-Pedersen and Olson 1978). Though such projects certainly provided a wealth of information about the physical remains of those particular vessels and established excellent methods for conducting underwater archaeological investigations, their significance to anthropological archaeology was often neglected. By the 1980s anthropologically minded maritime archaeologists began to realize the potential for shipwrecks to provide a greater range of information (Gould 1983). According to Gould (1997:380), "anthropological approaches encourage maritime archaeologists to look analytically at their results not only in relation to the specific details of maritime history but also to broader, social-scientific conclusions about human behavior in relation to marine environments."

Over the past 40 years, calls for more consideration of theoretical perspectives within the subdiscipline emanated from many parts of the world (e.g., Babits and Van Tilburg 1998; Blackman 2000; Gibbins 1990; Gibbins and Adams 2001; Gibbs 2006; Gould 1983, 1997, 2011; Hosty and Stuart 1994; McCarthy 1998; Potter 1990; Staniforth 2000; Veth 2006; Veth and

McCarthy 1999). In a journal article published at the turn of the twentieth century, Peter Veth and Michael McCarthy argued that a "vital component of any maritime archaeological reconstruction of the past must be clear and explicit statements of how specific nautical behaviors and belief systems can be reliably correlated with patterns in the material record" (1999:12). This statement highlights the perceived need for better informed interpretations for maritime archaeological data.

The initial demands for greater consideration with theoretical and methodological approaches led practitioners to infuse interpretations of wrecked ships, coastal landscapes, and other subjects covered by maritime archaeology with anthropological and archaeological theory. A synopsis of this can be found in Joe Flatman's "Cultural Biographies, Cognitive Landscapes and Dirty Old Bits of Boat: 'Theory' in Maritime Archaeology" (2003). This article provides an insightful discussion of not only the ways in which such approaches are applied to different maritime archaeological projects and site types but also offers several examples of how new approaches relating to shipboard societies might be explored. Some of the themes covered include gender, sexuality, class, social relations, hierarchy, labor, power, and discipline (Flatman 2003:145–150). One particularly relevant yet fleeting thought compares mining communities to shipboard communities: "The work by Pfaffenberger (1998) on the châines operatoires of different mining communities are of similar relevance, since mining communities are also controlled environments, including predominantly male populations, potentially dangerous and often clearly-delineated living and working environments[,] . . . strong social hierarchies, and a distinctive position in relation to wider communities" (Flatman 2003:147).

The interpretation of shipwreck sites specifically as industrial workplaces contributes a new approach to maritime archaeological research. As a research concept, the industrial workplace incorporates the technical and social components of industrial sites in its analyses and attempts to provide the most complete picture possible through documentary and archaeological sources. A holistic understanding of the relationship between the various components of sites can be gained by looking at associations between sources of raw materials, methods of processing, transportation, and the social context of production (Palmer and Neaverson 1998:4–5).

Conceptual Adaptation

As with all anthropological and archaeological research, identifying the key influences and thought processes that inform methodological and theoretical viewpoints is critical to understanding interpretations of the past. Though theoretical approaches pertaining to systems, critical archaeology, and landscapes are considered, themes and concepts are drawn specifically from those developed for industrial archaeology. Thus, three concepts—the industrial workplace, resource extraction, and industrial landscape—are discussed in relation to their manifestation in both industrial and maritime archaeological approaches.

The Industrial Workplace

The perceived need to consider the sociocultural elements of industrial operations results in archaeological sites being conceptualized and studied as interrelated industrial workplaces. In the simplest of terms, a workplace is any location where work is conducted. This description, however, fails to convey the fact that workplaces are multifaceted microcosms composed of many interrelated technical and social parts that require organization and order to successfully achieve goals. Obviously, the complexity of a workplace varies greatly, and factors affect it such as employee numbers, technical difficulty of work, or level of sophistication of the practices involved. Since the components of workplaces are particular to different industries, some archaeological studies aim at understanding them through structure interpretations using broad themes such as production (Riley 2005) or consumption and space (Mellor 2005) to create an umbrella under which all available data can be considered.

The desire to derive overarching concepts associated with issues related to industrialization and industrial society led to the proposal of a research framework for industrial archaeological sites that included the workplace (Palmer 2005). Covering issues addressed by industrial archaeologists over the past half century, this framework is structured as a list of nine broad research themes followed by supplementary questions that are intended to help investigations. The themes included in this framework are continuity and change; production and consumption; understanding the workplace; industrial settlement patterns; class, status, and identity; social control, paternalism, and philanthropy; use of scientific analysis in understanding significance of artefacts and industrial residues; historic landscape characterization; and international context of industrialization (Palmer 2005:16–17). Of particular interest is the theme "understanding the workplace," which is accompanied by questions pertaining to the ways in which issues like technological change and social control can be seen within the archaeological record. Although Marilyn Palmer's (2005:17) intent was to create a general "framework of inference" for understanding the social, economic, and technological meanings of the physical remains of past industrialism at archaeological sites, she was quick to point out that, as with the initial attempt at any set of guiding principles, it must be seen as a work in progress.

Robert B. Gordon and Patrick M. Malone first introduced the idea of a "maritime workplace" in their highly regarded volume *The Texture of Industry*, though provided little direction for ways in which it might be approached archaeologically. The reference to this concept is found within the section of their book titled "Industrial Workplaces," which is organized around *micro-geography* or "the spatial and functional relations within workplaces" (Gordon and Malone 1994:6). Included in their discussion is a brief mention of how shipboard processing of whale oil in the nineteenth century was an industrial experience. The fact that Gordon and Malone did not include a formula for investigating maritime workplaces is not surprising since doing so was not their intent. Instead, they simply identified shipboard whaling as a largely unmechanized, early American enterprise that might be studied archaeologically to learn more about the processes involved (Gordon and Malone 1994:228). Their reference to a whaleship as a maritime workplace, however, provides a starting point for understanding the structure of the workplaces that existed onboard pelagic whaling ships during the period under consideration.

Related to the notion of a ship as a maritime workplace are the concepts of *life aboard*, *shipboard communities*, and *shipboard society*, each of which are discussed within the subdiscipline of maritime archaeology (Dellino-Musgrave 2006; Muckelroy 1978; Redknap 1997). While these concepts capture the social and cultural aspects of life at sea, studies of them typically do not emphasize ships as workplaces or settings for industry. Further, very little research within maritime archaeology has focused on the community of a ship's crew; instead maritime historians have largely conducted that work (Norling 1996). As suggested by Flatman (2003:151), much work needs to be undertaken in maritime archaeology to explore "the nature of shipboard societies and their relationships to "mainstream" society, utilizing class, race, and engendered perceptions of social interaction."

Resource Extraction and the Sea

One approach to interpreting the remains of pelagic whaleships as industrial sites is to consider whaling as a marine extractive industry. Extractive industries have been the focus of archaeological inquiry for decades and an increasing body of literature has offered methods for their investigation and interpretation (e.g., Baxter 2002; Franzen 1992; Hardesty 1988, 2010; Knapp et al. 1998; Mate 2010, 2013; McGowan 2003). The term *extractive industry* incorporates all activities associated with the primary extraction of raw materials and includes mining precious and nonprecious metals, ore deposits, and fuel, as well as quarrying stone and forestry-related activities (Baxter 2002; Franzen 1992; McVarish 2008:287–323).

The archaeology of extractive industries, and in particular mining, has been an important part of industrial archaeological research. As with other types of industrial sites, early efforts to understand the remains of mining activities employed a technologically focused approach with a view toward preservation. Studies generally focused on the techniques employed, and physical remains were used to interpret how raw materials were obtained, processed, and prepared for transportation to other sites where they were refined in some way or used as fuel. In the case of remote mining districts or company towns, various types of housing were provided to offer an "improved standard of living" for workers (McVarish 2008:290). When included in early surveys, these facilities were simply described structurally and functionally with little or no consideration for the conditions experienced by the people that occupied them. Over time, however, practitioners realized that technology cannot be separated from people, their daily lives and human abilities, their ideologies and beliefs, and their capacity to negotiate complex social, economic, and political relationships within the industrial context (Childs and Kilpatrick 1993; Knapp 1998:18). This revelation resulted in research agendas that are more considerate of industrial culture and that attempt to explain sites as places where people performed work, rather than simply as the remnants of production.

Although no specific framework for extractive industrial site analysis has been formalized, archaeologists interpret such sites using general sociotechnical themes to explore the technology and overall experience of individuals working within the industry. Of these studies, perhaps no better example exists than Donald Hardesty's research into the processes by which extractive industries shaped the nineteenth-century cultural landscapes of Nevada, California, and Idaho. The archaeological and historical data analyzed in Hardesty's work are related to various sites of hard rock gold and silver mining conducted in Nevada from the latter part of the nineteenth and early twentieth centuries and are presented under the general themes of *technology* and *residential settlement* (Hardesty 1988:12). Hailed as the seminal work on frontier mining industries (Symonds and Cassella 2006:147), Hardesty's research provides a straightforward approach to documentary and archaeological records related to the industry (Hardesty 1988:ix).

Drawing on a maritime analogy to describe his approach, Hardesty characterizes the American mining frontier as "a network of islands colonized by miners" with each island representing a different source of raw materials (Delgado 2006:52; Hardesty 1988:ix, 1). He explains that unlike many of the mining sites recorded by industrial archaeologists, which include examples of surviving machinery, buildings, and landscape structures, the archaeological deposits found at many frontier sites are characterized simply by the remains of miner behavior. Thus, he states that his research is guided by an industrial archaeology that is defined as "the systematic problem-oriented study of the material remains of the workplace and the worker, in the context of the industrial revolution" (Hardesty 1988:17; Teague 1987:200).

Marine resource extraction is understood to be the removal of sea life for commercial purposes, and archaeological investigation of relevant activities include pearling (McGann 1999), fishing (Carter and Kenchington 1985; Raupp 2004), trepang fishing (Macknight 1976; Wesley et al. 2016), sealing (Anderson 2004; Henderson 1989), shell fishing (Botwick and McClane 2005; Shefi 2006), and turtling (Smith 2000). While these industries had similarities to whaling through the extraction of natural products from the marine environment, none required the vessel to be an industrial platform, nor did they require the length of voyage and intensity of production as did pelagic whaling.

Landscapes and Environmental Context

Landscape analysis is another approach that assisted in developing an understanding of the interrelationship between the lives of those employed in industrial mining activities, the technological aspects of the operations, and the environment. Mining sites were inextricably tied to the landscape and left unmistakable physical evidence of their existence. The use of landscapes within archaeological research provides not only a wide context within which to place sites and associated structures but also offers a mechanism for analyzing the social structuring of the physical (Hodder 1987; Newman 2001:100).

The concept of the industrial landscape was first popularized in the early 1980s by Barrie Trinder, and since then, it has become an essential tool for industrial archaeologists (Trinder 1982). Over the past 40 years, the number of surveys of industrial landscapes has increased, and, although the scope of those studies varies widely, the surveys commonly include elements such as buildings, machinery, pathways, and worker housing, as well as topographic features of the land and phenomenological qualities (McVarish 2008:373; Quivik 2000:56). The application of a landscape approach to extractive industries provides the ability to better understand the industry not only by its technology and workplaces—as defined by areas of site-specific activities—but how these were but one component in the strategy of a larger interconnected physical, social, and economic landscape.

The concepts of maritime cultural landscape (Ford 2011; Westerdahl 1992) and seascape (Cooney 2003; McKinnon et al. 2014) have been employed by archaeologists for many years to investigate maritime activities on land and water. While the maritime cultural landscape concept is generally used to describe changing coastal zones and human interaction within the land/sea interface, seascapes provide perspective for seafarers and their interaction with the open ocean far from land. When engaged in hunting activities, whaleships and their crews spent extensive periods at sea and only made relatively brief contact with land when fresh provisions were needed. Thus, the environmental context for pelagic whaling is largely that of the open ocean, and therefore, the concept of the seascape is applicable.

Changing Lenses

To better understand the industrial nature of pelagic whaleships, the aforementioned themes borrowed from industrial archaeology have been adapted to incorporate a maritime perspective. Much like Hardesty's (1988) approach to frontier mining sites, use of broad themes allows for a large amount of historical and archaeological data to be synthesized to provide a more complete picture of ships as technology, of industrial life onboard, and of ways in which the environment shaped and was shaped by maritime activities. By adjusting parameters to accommodate the confines of the ship and the hunting strategies employed by whaling captains, these themes become more focused on the seaborne elements of the early nineteenth-century pelagic whaling trade and on the wide-ranging nature of the whaling ships operating in the Pacific Ocean, which ultimately helps to define them as a maritime industry.

Before examining how early nineteenth-century pelagic whaleships functioned as industrial workplaces within the environment of the Pacific Ocean, it is necessary to understand them under the theme of maritime resource extraction. Due to the nature of hunting whales in often unchartered open ocean environments, the industrial success of a voyage also required a great deal of preplanning, decision-making, and outfitting to ensure that the ships were well-managed and self-sustaining entities capable of spending long periods of time at sea. And at the end of a voyage, there remained several activities related to postprocessing such as accounting, the settlement of debts, and the preparation of the ship for its next voyage. Thus, for the purpose of this discussion, the industrial system under which whaleships operated included three main stages: the precruise stage, the cruise operations stage, and the postcruise stage.

As with frontier mining, pelagic whaling required a high degree of organization, effort, and coordination at all stages to succeed in extracting raw materials and transporting them to markets. At its center, shipboard whaling involved a series of complex and interconnected processes conducted by a crew who lived and worked within the close confines of a relatively small platform for several years at a time. The industrial technologies employed on these ships were specific to whaling, and the ships themselves were only one technological component. Early nineteenth-century whaleships were unlike other contemporary vessels in that they were working platforms on which technological and social processes combined to allow for the extraction of raw materials from the marine environment. Much like the site complexes associated with primary resource extraction industries on land, commercial success hinged upon sets of interrelated and standardized practices specific to the industry. Thus, analysis of a pelagic whaling vessel as a resource extractive industry requires an understanding of the ship and how it provided the foundation on which whaling operations took place, as well as the gear needed to carry out specific processes involved in hunting and capturing sperm whales, processing blubber, storing whale oil, and transporting it.

Gordon and Malone's (1994) suggestion that pelagic whaleships were maritime workplaces provides an opportunity to further distinguish them from other types of commercial craft of the period. As stated previously, most commercial vessels of the nineteenth century can be viewed as maritime workplaces since their functionality depended on the use of specific tools and the management of labor through defined roles. The fact that pelagic whaleships operated not only under those conditions but also incorporated the equipment necessary to carry out resource extraction and processing sets them apart as floating industrial sites. Thus, the wrecks of early nineteenth-century pelagic whaleships can be analyzed and interpreted as maritime industrial workplaces.

Hardesty (2010:109) indicates that, when looking at settlement systems and mining site workplaces, the information "reflects the spatial arrangement of tools, operations, and social formations as well as the coordination of work within mining related socio-technical systems." As such, a better understanding of the whaleship as a sociotechnical system and workplace emerges. The individuals who operated and lived onboard early nineteenth-century whaleships provided the power for industrial processes and coexisted as a community aboard the vessel. This community can provide a focal point for understanding the industrial operations and relations associated with pelagic whaling. Although these communities were socially and spatially remote for long periods of time, they were linked to broader social and cultural networks by virtue of their role in supplying raw materials to a world system.

Much like remote mining communities and their associated workplaces, a maritime workplace represents "the domestic space of people who often were or are heterogeneous in character, of diverse origins, and drawn together by the need to work" (Knapp 1998:4). Order and efficiency were maintained onboard whaleships through a strictly defined hierarchy in the division of labor that consisted of officers, specialized craftsmen, and numerous general laborers. Whaleship crew members understood their positions within the chain of command and were tasked with specific duties and responsibilities relative to it; however, this awareness does not discount the active participation that resulted in various actions including rising through the ranks, desertion, and even mutiny. Though early pelagic whaling crews were generally composed of young men eager to make their careers as whalers, as the Pacific fishery grew, international market factors affected their composition (Brundage 1948). Over time, desertion and accidental death became commonplace and resulted in the need to replace crew members. Taking advantage of time spent replacing fresh provisions, many captains recruited from the Native populations of the Pacific islands they visited. Thus, the communities that existed onboard these ships were not stagnant in their makeup but changed over time.

The hierarchy that provided structure to the community onboard early nineteenth-century whaleships also manifested physically. The structure is reflected not only in the physical arrangement of the various industrial operations and working spaces but also in the accommodation offered. As with some merchant vessels of the time, personal space onboard whaleships existed at either end; average mariners were housed in sparse and cramped conditions in the front of the ship, known as the forecastle or fo'c'sle, while officers enjoyed better comforts in its more spacious after end. The physical separation and variable amount of space provided depending on hierarchy highlights the complexity of the seafaring experience between those that sailed on whaleships of this period.

The final theme to be adapted from industrial archaeological practice is that of the industrial seascape. Due to the mobile and fluid nature of the industry, this theme pertains primarily to the various hunting grounds and reprovisioning points visited throughout a voyage and the dangers, risks, and benefits of accessing them. Whaleship owners and agents did not simply send vessels out randomly to a given ocean for years on end in the hope of finding whales; instead, they made calculated decisions about where the vessels should operate based on collective knowledge of experienced captains. Hunting grounds were areas where whales were known to migrate, feed, mate, or calve, and though often separated by long distances, several potential hunting areas were accessible in different geographic regions (Finney 2010:46). Knowledge of these areas was invaluable and generally passed between the fraternity of whaleship captains before being added to printed charts. Information about particular grounds included data pertaining to relative population densities of different whale species and their seasonal movements; potential hazards (both charted and uncharted), such as shallow shoals and barely emergent reefs; seasonal weather patterns; and hydrographic information about currents and tides. Thus, for mariners the seascape represented the knowledge of an area obtained by experience and by the comprehensive review of all information at hand, including published charts and sailing guides, as well as the oral testimony of peers.

Although the industrial seascape in this volume pertains primarily to hunting grounds in the Pacific Ocean, it also incorporates the islands used by whalers for reprovisioning and for seeking refuge from storms. The positions of these were generally identified either through exploratory voyages or unexpected encounters with uncharted dangers. When sighted, captains recorded locations and ascribed names to such places. If a newly sighted island appeared to offer potential for resupplying a ship with fresh water and food, it was often reconnoitered to determine the availability of decent anchorages and whether safe approaches existed. Such investigations often produced positive information, and, over time, many well-known provisioning ports were established throughout the Pacific region. As with information about new hunting grounds, data relating to such places was initially shared among whaling captains but eventually found its way onto printed maps of the region. The long-term effect of these early encounters can be seen on modern maps, which include many remote and still uninhabited geographic features that bear the names of whaleships, their captains, whaling ports, or shipowners.

Conclusion

By the time American whaleships rounded Cape Horn to enter the Pacific Ocean, they operated under an effective system resulting from efforts begun in the second half of the eighteenth century to standardize technology and methods. This system allowed for the mass exploitation of sperm whales, which in turn fueled an ever-growing demand for whale oil in European markets and led to rapid expansion within the fishery. In much the same way that mining interests operating in the nineteenth-century American West explored the frontier in search of valuable ore deposits, whalers pushed further into the Pacific in search of new hunting grounds. The knowledge of regional geography and seasonal behaviors of their prey gleaned during the early years of Pacific whaling cruise allowed for refinement of the system and maximized profits.

Unlike other fisheries, pelagic whaling was industrial in scope, and the vessels employed were roving oil production platforms. Some of the less fortunate of these ships were cast upon reefs and atolls across the expanse of the Pacific, and their remains provide an opportunity to investigate the industrial system under which they operated. Drawing from analytical approaches used by industrial archaeologists provides a thematic perspective that has yet to be considered in maritime archaeology and allows for the bridging of the two subdisciplines.

2

The Rise and Prosperity of the American Whale Fishery

The origins of organized whale hunting can be traced over millennia in areas around the world. Likely beginning with scavenging whales that washed up on beaches, the bounty provided by their meat, bones, and oil eventually led to the development of trapping methods and active pursuit along the shore using small watercraft. The expansion of those activities to a commercial scale, however, is generally considered to be of European origin. Coinciding with rapid population growth in urban areas, the value of whale oil as a source of domestic lamp fuel led to increased market demand. Operations expanded until local whale populations were depleted, thus leading to hunting further from continental shores. In due course, whalers encountered the Atlantic Coast of North America and found it teeming with cetacean populations, which eventually gave rise to a global industry. This chapter provides a summary of human and whale interaction from early evidence of prehistoric hunting to the end of the "golden age" of commercial whaling in the mid-nineteenth century.

Prehistoric Evidence

In many parts of the world, opportunistic and targeted whaling was practiced since prehistoric times, when the oil, meat, and bone derived from stranded whales made them rich prizes (Sanger 1995:15). Often the result of severe weather, opportunistic strandings provided local inhabitants of nearby coastal regions with an unexpected windfall of food, fuel, and even building materials (Credlund 1989:46). Datable archaeological evidence of early whaling includes harpoons fashioned from bone found in Paleolithic cave sites along western European shores (Whipple 1979:43), as well as Neolithic rock art depicting whale hunting scenes from sites on the Korean peninsula (Lee 1984), areas of Russia (Lubanova 2007; Savelle and Kishigami 2013; Stoliar 2001), and Fennoscandia (Gjerde 2010). Furthermore, subsistence whaling practiced by Arctic cultures, such as those of the northwest coast of North America, has long provided an important source of energy (McCartney 1984:82; Taylor 1988:123). Though exactly how long Indigenous cultures engaged in targeted hunting of different cetacean species is unknown, the connection to whales remains significant as seen through their incorporation into belief systems that continue to this day.

Likewise, the exact origins of European whaling are uncertain. Archival evidence indicates that Scandinavian people in the Viking era (approximately AD 800 to 1066) hunted whales along the shore and developed methods that would later spread to other European cultures (Schokkenbroek 2008:26). An early example of this diffusion comes from an account documented by King Alfred of England, passed on to him by a Flemish explorer named Ohthere. In the account, Alfred describes a ninth-century method of taking whales by confining them in the fjords and bays along the coast of northern Norway before harpooning them (Lindquist 1993:24; Spence 1980:10). Other reports dating from the ninth to the thirteenth centuries include those of whaling by Normans, Icelanders, Germans, and Russians (Proulx 1986:9–13). Of all the European cultures known to have engaged in early whale fishing, however, it is the Basque people of the modern border region between France and Spain who are best known for whaling on a systematic, commercialized basis from the twelfth century onward (Hohman 1928:19).

Basque Whaling

As in other parts of the world, it is probable that Basque whaling began with people simply taking advantage of whales that became stranded near coastal villages. If it was fresh enough, meat from these animals provided sustenance, though the main product extracted was oil, which was used as fuel for heating and lighting (Ommaney 1971:71). Over time these products became important, and fishers began actively pursuing whales in shallow waters just offshore. The Basque found whales to be relatively timid creatures, which made hunting them with hand-thrown harpoons possible (Grenier et al. 2007; Proulx 1986:15; Scoresby 1820:11). The skills developed for efficiently harvesting whales allowed for great success, and soon, the oil became an important trade commodity. Thus, Basque commercial whaling was born. As more fishers engaged in this pursuit, regional governments involved themselves in the operation; records and documents dating to the eleventh century provide the earliest evidence of active hunting through a report of government levies on whale takes (Dégros 1940:162; Hawes 1924:16; Proulx 1986:15).

To exploit the offshore marine resources of the Bay of Biscay, methods and technologies for capturing whales were standardized. Among the most essential of these were vessels that could withstand the challenges of the hunt. To construct their whaleboats, Basque builders took advantage of the timber resources and high-quality iron ore found in the region. Since the targeted species, the North Atlantic right whale, was migratory (Kenney et al. 2001), shipbuilders did not construct boats solely for the chase; instead, they crafted sturdy boats that could be employed for fishing and seasonal whaling. This tradition of building multipurpose craft continued, regardless of size. And, as the quality of the vessels constructed became well-known to other European nations, the Basque people became famous for their ironworking and shipbuilding skills (de Zulueta 2000:261–271; Grenier 1988:69; Grenier et al. 2007).

The methods and technologies developed for hunting and capture during the early period of Basque whaling are significant, as they became the basis for those employed by other European and American whaling interests over the next 300 years. The Basque used stone towers as seasonal lookouts for whales and, once spotted, lit signal fires and pounded drums to alert waiting crews of up to 10 rowers, a harpooner, and a boatsteerer, who launched their boats from the shore and chased the whale until a harpoon could be deployed (Whipple 1979:44). Once a wounded whale became exhausted, it was given a final blow with a lance, and its floating, lifeless body was towed to shore for processing (Appleby 2008:24). The main produce of these whale hunts was whale oil, which burned much brighter than vegetable oils and was used to make other products such as cleansers, antifouling for ships, fabric protectants, and pharmaceuticals (Grenier 1988:69).

The profitability of Basque whaling operations depended not only on their ability to catch and process whales but also on their success getting the products to market and negotiating good prices (Hohman 1928:19). Eventually their hunting prowess, combined with climatic changes in the bay, led to a decrease in the number of whales frequenting the nearshore waters (Proulx 1986:18). Thus, the increasing demand in northern Europe for whale products led Basque whalers to venture ever farther from the Bay of Biscay in search of whales, and by the fifteenth century, they pioneered long distance voyaging in pursuit of their prey (Nash 2003:8). These voyages eventually led them to the coast of North America, where along the shores of Labrador they found rich stocks of both fish and whales. The region soon became an important location for both the whale and cod industries, as fisheries on the eastern side of the North Atlantic already exhibited noticeable depletion. Though the exact date for the first Basque whaleships entering the Labrador region remains uncertain, archaeological evidence indicates that by the mid-sixteenth century industrial settlements for processing whales and fish were established around Red Bay (Logan and Tuck 1990).

Over time, Basque whaling operations developed from a simple shore-based activity to one that included the use of large sailing vessels carrying small boats with which to pursue whales in the open ocean. The larger vessels were not purpose-built for whaling; instead, they were adaptations of existing types that could be used for whaling or fishing operations (Proulx 2007:43). When employed in the whale fishery, these ships served a dual purpose. First, they acted as "mother ships" that explored the area, launching boats when whales were sighted and transporting unprocessed blubber back to shore for rendering. Second, they operated as floating warehouses, storing oil until it was ready to be shipped back to Europe at the end of the season.

The methods employed at Red Bay were essentially the same as those described above; thus, if whales were sighted near the coastline, they continued to be taken by boats launched from shore. But if whales were encountered offshore, the small whale-chasing boats (*chalupas*), were launched from the mother ship as close to the prey as possible. Chalupas were double-ended and built for speed, but their lack of bow and stern reinforcement meant they were not intended to be towed by whales. Instead, it is more likely that harpoon lines would have been equipped with a wooden float known as a *drogue*, which slowed a whale's attempts to escape (Grenier 1988:79). Once dead, whale carcasses were either stripped at sea and the blubber stored on the ship for later processing on shore, or if close enough to shore, the entire carcass was towed in for flensing and processing. Regardless of when and where the blubber was removed, it was processed into oil onshore at established tryworks stations located near deep water access so that carcasses could be brought in as close as possible (Logan and Tuck 1990:68). The tryworks were a type of furnace in which large copper cauldrons boiled continuously when operations were underway. (Barkham 1984:516). When the oil was rendered, it was transferred to wooden casks and then stowed carefully in tiers onboard a warehouse ship. This operation was likely carried out as soon the barrels were filled, since the ship's hold provided the best possible protection from storms or other hazards (Proulx 2007:74). With the appropriate number of casks secured, the ship was prepared for the return journey, where the oil was transshipped and then sold at markets in Spain, France, and England (Proulx 2007:74–78).

The abundance of prey in the waters surrounding Labrador provided Basque whalers with great success. Archaeological and historical evidence pertaining to the region around the Strait of Belle Isle, known to them as Terranova, indicate that by the middle of the sixteenth century at least a dozen ports were in operation (Grenier 1988:69; Grenier et al. 2007). Over the next 50 years, the Basque business in Labrador developed into one of the world's most important whale fisheries (Ross 1985:1). Evidence suggests, however, that, by the end of the sixteenth century, it was waning due to the increasing scarcity of whale populations near shore (Aguilar 1986:196). Trouble finding prey compelled Basque whalers to explore the Arctic region, and eventually, they encountered the coasts of Newfoundland, Greenland, and Iceland. Soon they united their energies with Icelanders interested in the prospect of new commerce, and the business continued to grow (Scoresby 1820:17). In his well-known account of the history of the northern fisheries, Arctic whaling captain William Scoresby Jr. indicated that, by the end of the sixteenth century, 50 to 60 ships from various nations participated (Scoresby 1820:18). While all of this activity brought the Basque financial success and certainly added to the expanding market for whale products, it also attracted the attention of other European nations interested in establishing their own Arctic whale fisheries.

Spitsbergen, the Dutch, and the English Northern Fishery

By the 1500s, the search for the fabled Northwest Passage led English mariners into the unexplored seas to the north, where they found new lands with large stocks of whales in the surrounding seas and bays (Watson 2003:9). The most significant of these were the islands of Spitsbergen, part of the Svalbard archipelago located in the Arctic Sea. Due to the immense marine resources that existed around the coast, the issue of discovery and ownership of Spitsbergen became heavily contested between the English and Dutch. While the former claimed that Sir Hugh Willoughby's and Stephen Burrows's mid-sixteenth-century sightings of land in the general vicinity gave them rights to it, most observers agreed that it was Dutch explorer William Barentz who found Spitsbergen, as he had not only gone ashore but named many of the islands (Laing 1815:85–86). Though neither nation accepted the other's claim, England and Holland each noted the success the Basque experienced in Labrador, and by the late sixteenth century, both were keen to exploit Spitsbergen's cetacean populations to develop their own national whaling industries.

Marked population growth in Europe around 1600 led to exponential increases in revenue derived from whale products (Schokkenbroek 2008:26). Whale oil was particularly sought after by European nations not only for its illuminating properties but also for its use as a lubricant and wool cleanser in the textile industries (Francis 1991:55). Realizing economic potential for the whale fishery, several European nations organized whaling ventures. Though areas rich in whales were known to exist throughout the North Atlantic, the abundance of whales around the islands of Svalbard led most countries to establish stations there in the early decades of the seventeenth century. Archaeological materials indicate that these islands could have been visited or even inhabited as early as the third millennium BC (Chochorowski 1991:395), but it was not until European whaling activities of the seventeenth and eighteenth centuries that serious exploitation of the region occurred.

The two countries that became most involved in the early Arctic fishery were England and Holland. The English fishery began with attempts by the

Muscovy Company, which was granted an early monopoly on the fishery by the English government. In the latter part of the sixteenth century, they sent ships on unsuccessful voyages to the Gulf of St. Lawrence and around Iceland (Jenkins 1921:75). It was not until the early seventeenth century, however, that whaling commenced around Spitsbergen (Jackson 1978:5–7). In 1610, the English sent their first expedition to the region for the sole purpose of whale fishing (*North American Review* 1834:84), and the Dutch began whaling operations there in 1612 (Braat 1984:473). From the start, these two countries came into conflict; their mounting commercial and naval rivalry also translated into competition for dominance in the fishery (Whipple 1979:46). The problems between them led to the establishment of the Noordsche Compagnie in 1614, an association of independent enterprises that operated collectively under a charter from the Dutch government to organize and regulate whaling interests, as well as to provide protection using armed ships (Hacquebord 1987:20).

In the early years of bay whaling in Svalbard, the participating nations contended for access to the best ports and whaling grounds, which quickly became a legal issue (Arlov 1993:81). As the territorial disputes continued, division of the coasts and bays of Spitsbergen among the states engaged in whaling efforts became necessary. Although each country who previously had a vested interest in the fishery was offered sections of the island's coastline, the dominant English and Dutch interests were assigned the most productive areas (Leslie et al. 1850:348–349).

Hoping to prop up their then failing Terranova fishery, the Basque also set their sights on Spitsbergen. Conflicts with their competition in the Svalbard archipelago, combined with the effects of crippling and seemingly unending wars with Spain and France, ultimately collapsed their whaling operations (Proulx 1986:23). While Basque whalers did continue to take prey in the open sea, Spanish depredations in their home ports kept them from processing their catches, which annihilated the fishery by 1636. Thus, to carry on their whaling talents, Basque sailors contracted their services as captains and harpooners on foreign vessels (Noel 1809:692). Other nations engaged in the fishery were only too happy to employ Basques early on, for by doing so their inexperienced crews could acquire knowledge of the methods employed in the fishery directly from the Basque whalers who perfected them in Arctic seas along the North American coast. Unfortunately, as soon as these nations mastered the art of whaling, they refused to employ their trainers and even went as far as banning them from northern waters and threatening to sink Basque ships that ventured into the region (Whipple 1979:46).

Initial operations at Spitsbergen were conducted via a method that became widely known as shore whaling. Essentially, the same method employed by Basque whalers for centuries, shore whaling utilized small whaleboats launched from shore to intercept whales that were later processed on shore at rudimentary tryworks (Gibbs 2010:4). The boiled blubber produced a thick oil referred to as "train oil," a name that probably derives from the Dutch word *traan* meaning tear (Jenkins 1921:39). These methods proved successful, and the species targeted for this fishery was the Greenland right whale (*Balaena mysticetus*), also known as the bowhead. Since immense quantities of the seemingly tame bowhead were found in the region, and since taking other species proved quite dangerous with existing technology (Appleby 2008:25), they were taken by the hundreds and without an effort. Such destructive practices impacted whale behaviors and drove them first into remote bays and then far away from land into the open sea (*Dublin Penny Journal* 1836:347).

Thus, the rapid rate at which right whales were harvested made it necessary for ships to hunt offshore to meet quotas. This practice brought with it changes in the method of delivering blubber for processing; the increased distance from shore saw the use of another Basque method known as "bay whaling." This technique involved cutting the whale's blubber into small pieces that were then stored in casks on the deck or in the hold of the ship. Once full, the ship returned to shore where the blubber was processed at stations that were used year after year. In time, the pressure of bay whaling on populations pushed them even farther offshore and gave way to open sea whaling. Operated in the same way as bay whaling, open sea whaling required longer voyages farther out to sea and was far riskier due to dangerous Arctic sea conditions and the chance of plunder by ships of other countries—especially England (*North American Review* 1834:87).

Regardless of the fact that the English supposedly received the best part of the island in the earlier land division, by the early 1620s there was a noticeable decline in its involvement in the fishery. Among the main factors contributing to this decline was profitability, which was greatly affected by the massive increase in oil imports. This influx resulted in low priced oil in the English market and in turn led to less investment by shipowners (Proulx 1986:30). Another major problem faced by the Muscovy Company was poor internal management of the fishery; crew structure, wages, ship maintenance, and outfitting costs were all issues that greatly affected operations during this time (Proulx 1986:30). The shift from shore whaling to bay whaling to open sea whaling was another factor that led to the decline of the English fishery. The ever-expanding number of Dutch ships had a direct effect on whale populations throughout the region, which was in turn reflected in diminishing English productivity (Jackson 1978:25–26). Ultimately, though efforts were made to infuse the English fishery, the problems that plagued it proved too great, and from the end of the first quarter of the seventeenth century, it languished.

In the meantime, Amsterdam became a major European market for oil, and, as such, the Dutch pursued their whale fishery with such vigor that they soon dominated (Israel 1990; Scoresby 1820:41). Unlike the English, the Dutch invested immense capital in their whale fisheries, which soon became the source of national wealth. They also established the large settlement of Smeerenberg at the northwest corner of Spitsbergen to serve as a blubber processing station and supply outlet for their ships (Hyde 1874:127). The establishment of this outpost proved to be a lucrative investment, and by the heyday of the settlement, up to 200 ships visited it during the season (Spence 1980:32). Thus, the development of this infrastructure, the quality and experience of the Dutch sailors, the economic organization of their business, and the low cost of their ships were all factors that led to their virtual control of the northern fishery for much of the early to mid-seventeenth century (Braat 1984:476).

Thirty years of intensive whaling at Spitsbergen drastically reduced populations, and it was necessary to move operations west to the Greenland Sea, Davis Strait, and Baffin Bay (Isachsen 1929:388). As the ships cruised farther away from Spitsbergen, stopping at Smeerenberg became unnecessary, and the settlement flagged. The fact that most whaling was by that point carried out in the open sea meant that it became impossible to control; thus, the number of interlopers, or non-company-owned ships, involved in the industry grew at an incredible rate (Hacquebord 1987:143).

As a result of the interloper problem, the monopoly granted to the Noordsche Compagnie was removed in 1642 (Tower 1907:17). At the same time, changes in the Arctic climate saw the return of large numbers of whales around the Spitsbergen coast, and the fishery shifted its focus back there. Arctic archaeologist and whaling historian Louwrens Hacquebord describes the large numbers of whalers flocking to the region during this period as having taken "the character of a gold rush, mainly concentrated on the Spitsbergen hunting grounds" (1984:144). For many years, Dutch whaling continued to flourish, with an increasing number of ships visiting the grounds; between the years 1660 and 1670, the fleet comprised at least 400 ships annually (Scoresby 1820:56).

The inevitable consequence of the relentless hunting was a decline in the number of whales on the coast, which pushed the dwindling population among the ice floes. The dangers encountered in these areas resulted in unsuccessful voyages, the loss of many vessels, and eventually to the decline of the Spitsbergen fishery (Tower 1907:17). These risks in turn led to a search for new grounds. In 1719, Dutch whaleships first sailed into the Davis Strait and established a fishery that would remain active for the next 200 years (Clark 1887:194). Though this new fishery helped Dutch whaling to prosper for a time, the effects of external factors, such as constant war with other European nations, took their toll. Slowly the number of ships in the fleet decreased and by the last quarter of the eighteenth century Dutch superiority in the fishery waned (Tower 1907:17–18).

English whaling did not cease altogether during the seventeenth century, though its involvement was miniscule in comparison to the Dutch. In time, however, the sad condition of the fishery drew the attention of the British government. Officials not only realized its potential economic significance but also "saw its importance as a nursery for hardy mariners, as offering employment for a great number of ships, while the requisite equipments would require the co-operation of a number of artisans, tradesmen, and laborers" (Scoresby 1820:57). Thus, in an effort to help rebuild English whaling at this time, Parliament deemed it necessary to pass an act to stimulate the industry (Tower 1907:15). This initiative took the form of heavy duties on all whale products imported into Britain by foreign ships—including their own colonies. On the other hand, English-built whaleships equipped by mostly English crews that left from, or were returning to, English ports were exempt (Schokkenbroek 2008:36). Though these conditions did attract some interest from private firms between 1672 and 1697 (Scoresby 1820:58), overall they failed to revive the trade, and no serious attempt at whaling was made for the next quarter of a century (Jackson 1978:39).

By the 1720s, English demand for whale oil was high. Thus, the discovery of the rich Davis Strait whaling ground rekindled interest in the Arctic fishery (Francis 1991:55). In 1725, the London-based South Sea Company began investing heavily in the implementation of a seasonal fishery in the area, but lack of success resulted in financial losses and led to its abandonment in 1732 (Leslie 1850:351–352). Petitions from the company to the government to subsidize their speculation in the fishery, however, came to fruition; in that same year, the government implemented a bounty of 20 shillings per ton on all British whaling ships exceeding 200 tons (Scoresby 1820:72). The bounty had some positive effects and slowly encouraged more ships to outfit whaling voyages, but it was not until it was doubled that the industry began to revive (Proulx 1986:32). The Bounty Act of 1749 immediately resulted in greater investment in the fishery and induced many seaport towns to outfit vessels for whaling (Jenkins 1921:185). Soon, northern English ports, such as Hull and Whitby, as well as others in Scotland, including Aberdeen and Dundee, became involved, and their importance grew quickly in the following decades. Thus, the 1749 increase in the bounty was a boon that revived British whaling, gave birth to the Scottish fishery (Watson 2003:12), and ultimately provided the impetus for British dominance in the Arctic by the beginning of the nineteenth century.

The New England Fishery to 1750

At the same time European nations competed for the prime hunting areas around Spitsbergen, England established colonies in North America. With new colonies came the prospect of exploiting natural resources in the region.

Reports from the ships that initially explored the region mentioned immense quantities of whales and abundant stocks of fish in its bays and nearshore waters. Thus, fisheries were considered by the English as one of the main commercial enterprises to be undertaken. An example of the perceived importance of these activities was seen when Captain John Smith set out for the colonies in 1614 and carried a Crown permit to fish for whales (Ashley 1926:28). His return to London only six months later with a cargo of dried fish, furs, and whale oil confirmed the potential for wealth to be obtained from these untapped seas (Dow 1925:5–6).

The colonists were not the first to realize the value of taking whales along this coast; Native Americans of the region considered whales a source of sustenance. Coastal Native Americans regularly took advantage of sick, injured, and dead whales stranded on the shore (Whipple 1979:47). When actively hunting, they employed dugout canoes to pursue their prey and harpooned them with stone-tipped wooden spears equipped with wooden drogues that slowed the whales down (Bailey 1953:83). Observing their whaling practices, colonists noted that the flukes and fins were the only parts taken and that carcasses were left for the birds. As they were only interested in the blubber from which they could obtain much sought-after oil, this situation proved perfect. To maintain good relations with their Indigenous neighbors, they proposed a situation that would benefit all—when whales were found or taken, each group offered the discarded pieces to the other (Bailey 1953:83). Colonists at Southampton, New York, also gave them harping irons, or metal-tipped harpoons, to increase the effectiveness of Native American whale hunts (Palmer 1959:3).

The frequency of coastal whale stranding meant that settlers could easily undertake drift whaling, which involved simply processing carcasses of whales that died of natural causes and drifted ashore and then storing whale oil for shipment to markets abroad (Fairburn 1945:977). Such opportunistic whaling was initially seen as a financial windfall; however, issues of ownership and disposal of drift whales appear to have attracted much attention in both the Plymouth and Massachusetts Bay Colonies (Tower 1907:20). The governments of each colony, as well as those on Long Island, decreed whales could not be the exclusive property of the person or persons finding them. Instead, they belonged to the public, and thus, the township where the drift whale came ashore was entitled to an equal share of the oil and products salvaged (Fairburn 1945:977). The profitability of these early activities and the knowledge that whales were easily found just offshore soon led to the establishment of shore whaling enterprises (Allen 1928:340).

As drift whaling was opportunistic, obviously the type of whale that washed up was of no concern to the colonists or Native Americans. Once whales were actively sought by the shore fisheries, however, two main species were targeted:

the North Atlantic right whale, which frequented the New England coast from October to June; and the humpback (*Megaptera novaeangliae*), present in the summer and fall (Ashley 1926:30). Of these, the North Atlantic right whale was the main object of the hunt because they produced large quantities of oil and whalebone (Reeves and Mitchell 1986:201). It appears that the term *right* was given to them simply because they were the right kind to take due to their natural tendency to float after being killed (Allen 1916:171; Braginton-Smith and Oliver 2008:19).

The exact location of the first shore whaling activities on the New England coast is contested. While some authors have attempted to identify a particular time and place, it is likely that the first whales were taken by colonists who had some understanding of, and possibly participated in, English or Dutch whaling in the Arctic prior to their arrival in the New World. The earliest records of organized whaling operations come from Long Island, where, by the late 1640s, "whaling companies were formed, neighborhood lookout stations were posted, and small craft put out from shore on cruises that lasted one or two weeks, making shore each night" (Ashley 1926:29–30; Starbuck 1878:10). With the success of these endeavors, shore whaling spread along the coast, and both the number of boats engaged in the business and the number of whales towed ashore increased (Dolin 2007:47).

The main method employed in early shore whaling was a technique that came to be called inshore whaling and involved cooperative efforts only slightly different from those developed by Basque whalers (Davis et al. 1997:43). As mentioned previously, lookouts were stationed along the coast for the purpose of scanning the ocean for spouting whales. These structures consisted of tall spars fitted with wooden pegs for climbing, topped with elevated seats, and covered with thatched roof shelters that were open to the sea (Forman 1966:31). When whales were spotted, the lookout alerted the companies, who rowed out in whaleboats with a complement of six crew members (Dolin 2007:49)—many of whom were Native Americans paid for their skills as harpooners. The boats developed for shore whaling were constructed of native cedar (Cupressaceae) and described as "light, graceful and easy to manage, [and] were exceptionally strong" (Bailey 1953:83). When a boat was close enough to a whale, it was struck with a harpoon attached to a line and allowed to run; when exhausted, it was then killed by a blow from a lance. Once dead, the whale was towed to shore, and, with the assistance of a capstan-like machine called a crab, the blubber was removed and either boiled in metal pots close to the beach or taken to one of several more distant tryhouses for rendering (Hawes 1924:73). Such tryhouses were semipermanent facilities that offered better conditions where rendering could be done in a more orderly fashion (Weiss et al. 1974:79).

Shore whaling proved profitable, and soon, it became an economic mainstay for the communities of Long Island and the New England coast (Francis 1991:45). The proceeds from these activities did not go unnoticed by colonial authorities who sought to regulate the new businesses by levying duties. For example, in 1684 an act was passed that "layed a duty of 10 percent on all oil and bone exported from New York ports to any outside ports except directly England or to the West Indies" (Starbuck 1878:15; Tower 1907:23). Despite these and other controls, shore whaling continued to grow, and coastal communities profited. The increased number of whales being taken, however, led to a noticeable depletion of right whale stocks by the beginning of the eighteenth century (Braginton-Smith and Oliver 2008:75).

Eventually the success of shore whaling on the mainland led the inhabitants of the small offshore island of Nantucket to adopt the trade (Figure 2.1). European colonization of Nantucket occurred around 1650 when a group of Quakers fled persecution on the mainland. "The character of the island and its situation far out in the ocean, its poor soil and the number of whales along its shores, all proved inducement to the Nantucketers to follow the sea as a calling" (State Street Trust and Walton Advertising and Printing 1915:19). Though whales could easily be found along their shores, for some time the islanders only took advantage of drift whales that washed up—many of which were previously harpooned by their neighbors on the mainland (Ashley 1926:30). In time, they attempted to kill whales that ventured into the bays and harbors and quickly realized the economic potential of a well-organized whale fishery. To instruct them in the best methods for taking whales and extracting oil, the Nantucketers enlisted the help of a Cape Cod whaler named Ichabod Paddock. Paddock moved his family to the island and quickly took up his new role by dividing the island's southern coastline into four sections and establishing lookouts operated by crews of six people. Whaling became a community affair as all shared in the expense, the work, and the products and profits (Fairburn 1945:981). Nantucket's Indigenous inhabitants played a key role in early shore whaling by allowing the establishment of whaling stations in prime coastal areas and undertaking all aspects of the fishery, often through coercion (Dolin 2007:69; Shoemaker 2015:10). Due to the diligence of the islanders and the cooperative nature of the fishery, Nantucket achieved such success that by the turn of the eighteenth century, it became the leading whaling port in the English colonies (Proulx 1986:60).

Over time, the numbers of whales frequenting the coasts of New England dwindled. Realizing that large stocks of whales could be found offshore, whalers extended their operations farther out to sea. Thus, by the end of the seventeenth century, boat whaling had begun in New England. Boat whaling did away with the need for lookouts by sending boats out to actively hunt

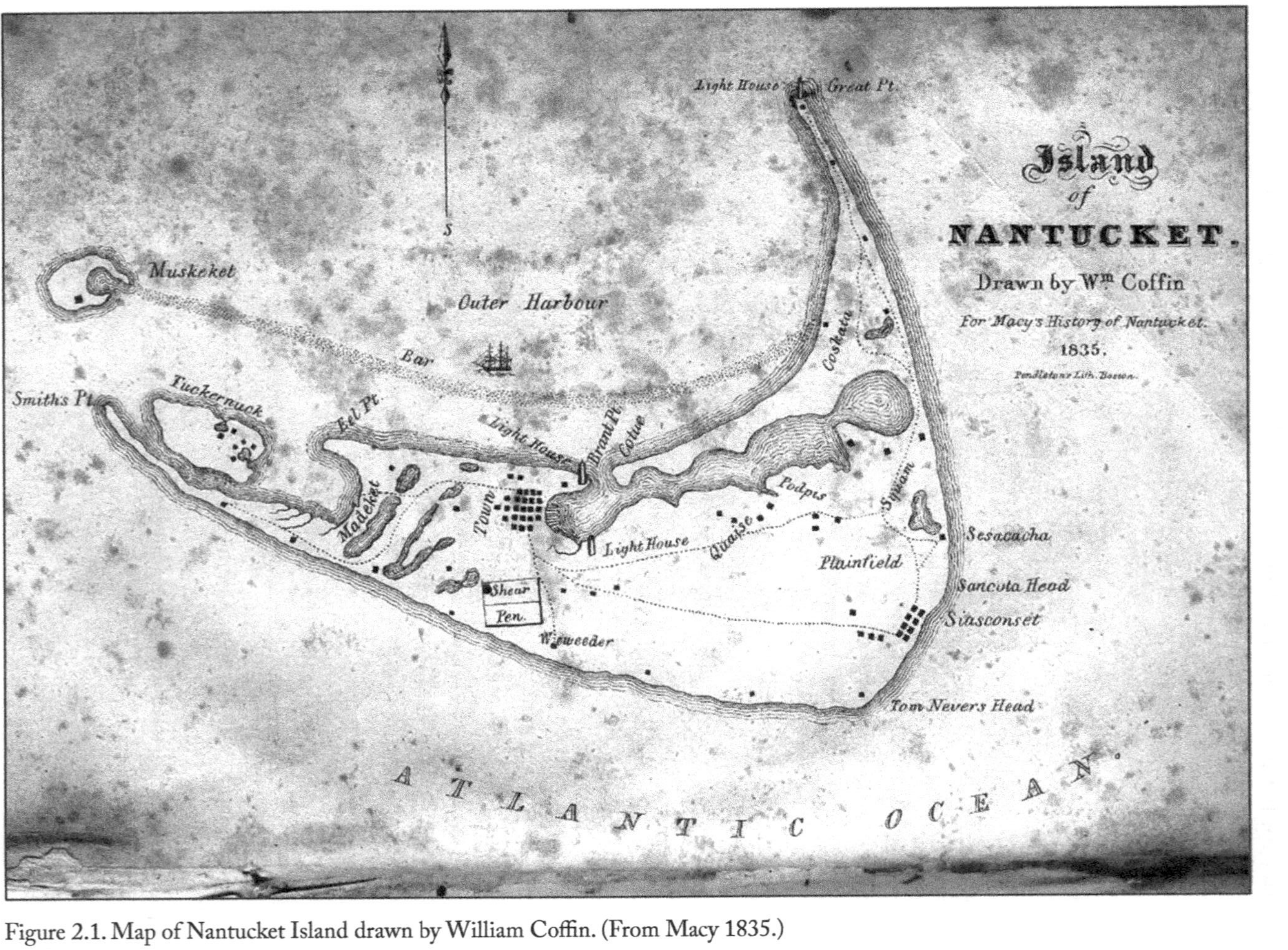

Figure 2.1. Map of Nantucket Island drawn by William Coffin. (From Macy 1835.)

whales in nearshore waters. This involved outfitting small boats with provisions to last a few weeks and then cruising in search of prey. When a whale was taken, it was towed to the nearest shore and flensed before the blubber was packed into barrels and taken back to the home port to be processed while the whalers resupplied and returned to the hunt (Davis et al. 1997:35). By the early 1700s, Nantucketers were taking their sloops on voyages of 60 to 80 kilometers (km) out to sea, well beyond the shoals and sight of land (Stackpole 1953:24).

Perhaps the most significant of these early boat whaling voyages was that of Christopher Hussey around the year 1712. Although some authors have disputed the exact date and accuracy of this event, it represents the first known instance of a sperm whale (*Physeter macrocephalus*) being actively taken at sea (Starbuck 1878:20, 768–769). According to Nantucket's first historian, Obed Macy (1835:36), Captain Hussey was "hunting Right whales when a strong northerly wind blew him some distance offshore, where he fell in with a school of that species of whales [e.g., sperm whales], and killed one and brought it home." The sperm whale was known to the people of New England, as carcasses had been found stranded along their beaches (Figure 2.2). As such, they were aware that the oil from this species was of superior quality and that their heads contained a reservoir of a waxy substance known as spermaceti. The reason that whalers had not actively pursued them was simply that there were plenty of right whales close to shore (Spence 1980:43). The commercial value of sperm oil, however, was undisputed, for it was the highest quality known (Stackpole 1953:28). Thus, the significance of Hussey's voyage was that it proved these whales could be taken in open water and that it inadvertently marked the advent of offshore whaling or pelagic whaling.

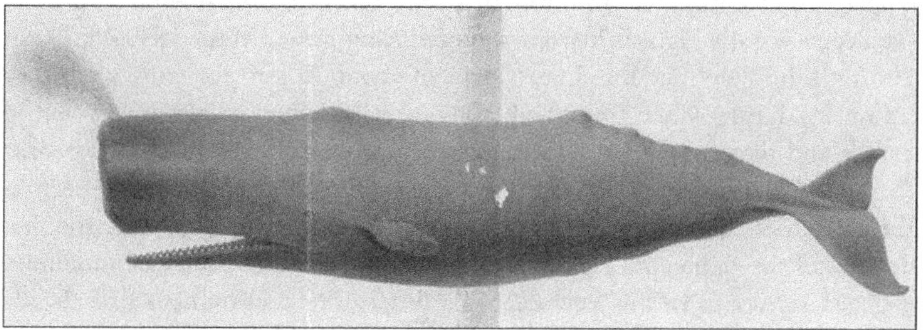

Figure 2.2. Illustration of a sperm whale (*Physeter macrocephalus*) by Charles Scammon. (From Scammon 1874.)

Offshore whaling required replacing small boats used in inshore and near-shore fisheries with much larger vessels, such as sloops, schooners, and brigs. When sloops were used, whales continued to be directly hunted from them; however, the larger schooners and brigs were operated by crews of up to 13 who attacked whales from two whaleboats, which were carried onboard (Davis et al. 1997:35). The larger size of the vessels was necessary to hunt the much bigger sperm whale because though undoubtedly sperm whales were a richer reward than the right whale they were also more dangerous (Whipple 1979:50).

Though sperm whales inhabit all of the world's oceans, Nantucket whalers first found them in the warm waters of the Gulf Stream that passed offshore of their island home. Their size, their unusually large head, and their behavior enabled them to be easily distinguished from other whale species. Feeding mainly on squid that live in the ocean's depths, they often make prolonged dives to feed. They also have a strong schooling instinct, forming schools of females and young males, while older adult males are solitary (Gosho et al. 1984:54). The carcasses of sperm whales yielded several products including sperm oil, which burned with a very bright odorless flame (Spence 1980:47); spermaceti, a waxy substance found inside their heads and used to produce the finest of all candles, as well as a lubricant for very delicate instruments; the fine-grained ivory teeth that were put to many uses such as carving and decoration; and the treasured ambergris, a by-product of an infection in the whale's intestine that was used as a fixative for high-priced perfumes and, thus, was literally worth its weight in gold (Sanderson 1956:211–212).

Due to higher prices paid for sperm whale products and decreasing numbers of whales along the coast, the deep-sea fishery soon eclipsed shore and boat whaling. Though Nantucket shore whalers carried on as late as 1760, the fishery is said to have reached its peak around 1726 (Macy 1835:46; Tower 1907:27). The profits that Hussey realized from his legendary voyage quickly led Nantucketers to outfit trips into the wider ocean—referred to as "ye deep"—to distinguish it from shore whaling areas (Macy 1835:36; Dolin 2007:90). Initially they fitted out sloops of about 30 tons for cruises of up to six weeks, during which time the blubber obtained from whales was stored in barrels and then brought back to shore for processing at established tryworks (Macy 1835:46; Tower 1907:27). The success of these activities can be seen in the growth of the industry and the number of vessels employed over the first decades of the eighteenth century. In 1715, the Nantucket whaling community engaged six sloops in this new deep-sea fishery, but that number had already grown to 25 vessels ranging from 30 to 50 tons by 1730 (Tower 1907:27).

The scope of the offshore whaling voyages quickly expanded. Bound only by the cargo capacity of their vessels and the provisions aboard, whalers from various New England ports cruised both north and south in search of prey.

Historian Ivan T. Sanderson (1956:212–213) succinctly describes the seasonal rhythm as "the practice was for the whalers to fit out during winter, then go south in the spring—first to the Carolina coast, thence via the Bahamas either into the Gulf of Mexico and the Caribbean or down the outside of the West Indies, then east to the Azores, from there to the Cape Verdes and the west coast of Africa. . . . The fleet returned to Nantucket in July, refitted, and then sailed again for the Grand Banks and Davis Strait to the north." Though sperm oil was in higher demand, American whalers did not discriminate when they encountered whales of any sort; instead, they simply kept blubber of different species separate. Other towns in the region, such as Salem, Massachusetts, and New London, Connecticut, also outfitted vessels for offshore whaling and enjoyed some success (Decker 1974; Robotti 1962). Nantucket, however, took the early lead and would become the center for American whaling until the early nineteenth century (Ellis 1991:143).

The colonial market for whale oil was based in Boston, where merchants controlled the shipping of products to foreign markets. The success of offshore whaling and the production of such large quantities of oil soon glutted the colonial market and forced prices down. Though London was the principal port for the New England trade, Amsterdam and other European ports claimed a share of the oil shipped through Boston (Dow 1925:24). Realizing that colonial oil was in constant demand on the other side of the Atlantic resulting from issues with the Spitsbergen fisheries, Nantucket whalers decided that a direct contract with foreign markets would be more beneficial. Though there is evidence of a shipment from Nantucket to the London market as early as 1720, it was not until the middle of the eighteenth century that a direct trade was established in which whale oil was delivered, and articles required for the whaling business, such as iron, hardware, hemp, sail cloth, and many other goods, were brought back to the colony (Dow 1925:24–26). By setting up this trade, Nantucketers bypassed the fees they had paid previously to Boston merchants on both ends.

In the first half of the eighteenth century, several improvements were made to the methods and technologies employed in offshore whaling. These advances included the practice of fastening the boat to the whale by means of a line attached to the harpoon, the evolution of the versatile whaleboats used to chase prey, better methods for stripping blubber from carcasses, and the development of higher-quality whaling craft and gear, such as harpoons, lances, line, and cutting spades (Hohman 1928:27–28). By this time, larger vessels were used as mother ships and floating warehouses; as such, they not only increased in size and tonnage but also underwent gradual changes in rig (Hohman 1928:27–28). While all these changes increased the range and certainly the productivity of the fishery, progress in the industry was still restricted. Limitations included the small size and relatively small cargo capacity of sloops and schooners employed as mother ships and the need for low temperatures to preserve raw materials, which restricted them to seasonal operations. Coupled with the growing need to make longer voyages to more distant grounds, these constraints led to a revolutionary innovation: the transfer of tryworks from the shore to the deck of the whaleship (Davis et al. 1997:36).

Installing tryworks on decks of whaling vessels marked a turning point in the industry. The ability to process whales wherever they were taken had major advantages over offshore methods. First, it lengthened voyages, which allowed for an increased number of whales to be taken and in turn produced greater profits. Second, it allowed whalers to move out of the northern hunting grounds to pursue sperm whales in the warmer waters they preferred. Several subsidiary effects of the application of shipboard tryworks included increases in the size of the vessels employed in order to provide greater oil storage capacity and increased seaworthiness, a subsequent change in the rig of the vessels, changes to their outfitting, and changes to labor requirements for pelagic whaling, including increased numbers of crew, clearly defined roles, and the institution of a wage system.

Though onboard tryworks resulted in access to an unprecedented supply of sperm whales, it was the development of a method for separating the spermaceti that greatly increased their value (Bockstoce 1984:529). The originator of this procedure cannot be identified with certainty, but it is apparent that the person had some knowledge of the process of tallow candle manufacture (Kugler 1980:5). Spermaceti is contained in a void in the sperm whale's head called the case. When exposed to air, it becomes a waxy substance that can be used to make high-quality candles (Stackpole 1953:28). The process of separating spermaceti involved several steps: first, the junk (the unprocessed matter) was heated in a large copper vat to remove impurities; then it cooled and congealed before being bagged in woolen sacks; finally, it was squeezed using a screw press (Ellis 1991:144). The oil obtained was of extremely high quality and was recognized as a far better illuminant than that procured from right or humpback whales, and from that point on, oil was sold in two classes, sperm or whale, with each fetching a different price (Kugler 1980:5).

Together these two technological innovations had a phenomenal effect on the market for whale products and the expansion of the fishery. The addition of processing blubber into oil greatly extended the lengths of the cruises, which in turn required the creation of a system that allowed vessels the freedom to roam the seas until their holds were full. From the types and sizes of vessels chosen and their outfitting prior to departure, to the taking of whales and stowage of oil, to the payment of wages upon return to their home port, all aspects of the trade were affected under this new system. Because it was the whalers of New England who developed the system, it is commonly referred to as "American-style whaling" (Davis et al. 1997:37). The introduction of this new system led to the ability to maintain cruises for years at a time, which in turn allowed American whalers to explore far beyond known grounds, and the resulting financial gains had a profound influence on their world.

Revolutionary Technology to Revolutionary War

By the middle of the eighteenth century, urban development led to a great social transformation in cities across the world—street lighting (Dolin 2007:121). On both sides of the Atlantic, whale-oil lamps lit homes, churches, factories, and shops; and oil was used for lubricating machines, finishing leather goods and woolen clothes, and making soft soap, varnishes, and paints (Hawes 1924:81). As a result, whale products were by this time one of the principal exports of the Massachusetts Bay Colony to England (Dow 1925:26). While whaleships continued to take whatever species they encountered, the increasing prices paid for sperm oil meant that they became the main targets. Historian Richard Kugler (1980:6) highlights the role of New Englanders in developing this fishery by pointing out that "sperm whaling quickly became an American preserve, exclusively so up to the time of the Revolution and largely so thereafter."

Some consider the period from 1750 to 1784 to be the most important era in the history of the whale fishery (Starbuck 1878:36). During this time, the character of the American deep-sea whaler emerged, as those involved quickly became experienced mariners with a growing knowledge of the hazards of both whaling practices and the industry's financial success (Stackpole 1953:47). As more ports along the New England coast outfitted vessels for extended whaling cruises, the boundaries for hunting grounds were pushed ever farther offshore. In addition to the dangers of this new style of whaling, English foreign affairs meant that American whalers faced the constant threat of attack by Spanish and French privateers and pirates who roamed the Atlantic Ocean (Starbuck 1878:56; Tower 1907:33).

The influx of colonial oil into the markets of London and other European ports was welcomed enthusiastically due to the almost complete failure of British Arctic whaling. The end of the French and Indian War in 1763 saw France concede all claim to Canadian lands to England; as a result, New England whalers eagerly moved into the Arctic (Ellis 1991:145). Around the same time, the British sought new sources of revenue and established a bounty system that favored English ships and their national fishery. The bounty system was extremely restrictive to the colonists, and by 1765, more discriminatory legislation put the colonial fisheries at a severe disadvantage by prohibiting their vessels from taking products to any non-English markets and by limiting the waters in which they could operate (Ashley 1926:32). Though seen as unjust, there was little action to be taken aside from petitioning Parliament, which was done by both New England merchants and the London firms involved in the colonial oil trade (Dow 1925:26). Ultimately these sanctions were only partially successful; while they did drive some New England whalers from the Arctic grounds, they failed to impede the growth of the colonial industry. Instead, the American whaling fleets turned their attention to the south (Francis 1991:61).

Whaling voyages by this time extended in length from weeks or months away from home to sometimes up to a year. These extended cruises led to further geographic expansion, and by 1767, 50 New England whaling vessels explored far southern waters (Fairburn 1945:985). A major scientific discovery during this early period of exploration was the identification of the Gulf Stream. Nantucket whalers were the first to gain knowledge of this powerful north and eastward flowing current, and one of them eventually passed the information on to Benjamin Franklin, who in 1786 published the first map (Figure 2.3) charting its path (Whipple 1979:54). The Gulf Stream abounded with sperm whale food, and, by hunting along its edges, rich new grounds were identified, ranging from Hatteras, off the coast of modern-day North Carolina, to the Banks of Newfoundland, east across the Atlantic Ocean, and south to the coast of Africa (Fairburn 1945:984–985).

Figure 2.3. Benjamin Franklin's map of the Gulf Stream. (From Franklin 1786.)

Also encouraged by reports of merchants engaged in the West Indies trade, whalers ventured farther south where they found their prey in abundance around the Bahamas in the Caribbean. Exploration of the Atlantic region continued, and by the mid-1770s, American whalers were hunting off the coast of West Africa, around the Azores, and off the coast of Brazil (Robotti 1962; Tower 1907). The hunting grounds identified along the South American coast, including those around the Falklands and other South Atlantic islands, would prove to be important for both American and revived British fleets. Whaling in this region was not introduced by American or British whalers; instead, accounts of Portuguese explorers describe the presence of whales in shallow waters off the Brazilian coast, which initiated a limited shore-based fishery by the seventeenth century (Alden 1964:270). Basque sailors introduced shore whaling techniques to the region, and by the mid-eighteenth century, shore processing facilities dotted many parts of the coast. The involvement of American, British, and French offshore whaling interests in the region, however, greatly reduced the whale populations and resulted in the decline of the shore fishery off Brazil (Richards 1994:6). Of all the new offshore whaling grounds identified in the exploratory period prior to the American Revolution, none proved as important as the Brazil Banks over the next decades.

The American Revolution, 1775–1783

Despite the British intrusion, the lead-up to the American Revolution saw the whaling industry of New England enjoy healthy progress and profit. Evidence of this success can be measured by the fact that in the year 1775, the Quakers of Nantucket and New Bedford had around 300 vessels engaged in the fishery (Bockstoce 1984:529). Other New England ports seeking to gain a foothold also fitted out 50 to 60 vessels. These included Wellfleet, Martha's Vineyard, Barnstable, Falmouth, Swansey, and Boston in Massachusetts; Newport, Providence, Warren, and Tiverton in Rhode Island; New London, Connecticut; and Sag Harbor on Long Island, New York (Dow 1925:40).

As most whaling historians indicate, the American Revolution nearly destroyed the colonial industry. In "The Whale Oil Trade 1750–1775," whaling historian Richard Kugler notes that the war not only shut down the industry but ended a phase of critical change in it. Referring to the quarter century leading up to the war, Kugler (1980:23) suggests it was "a time of expansion, of innovations in technology and management, and of rivalries and problems not completely resolved before the conflict closed down the fishery."

The threat of conflict between the colonies and the British government was ever present in the early to mid-1770s. Dissatisfaction was felt throughout the colonies over taxation and control. Sources of discontent included not only the duties imposed on the whaling fleet that fished in the Arctic but others that restricted colonial commerce. Ultimately, the commercial instability escalated to the point of violence in April 1775 at the Battles of Lexington and Concord (Jackson 1978:66). This unrest proved devastating to New England's fisheries, as they became the initial focus of England's attempts to repress the colonies (Tower 1907:37).

Of all New England ports, the most successful at sperm whaling was Nantucket. This success can be attributed to several factors, including its location off the coast of the mainland, which positioned them closer to the hunting grounds; the good business sense and determination of the Quaker whalers; and the system of community investment in voyages. All these features combined to create an island economy that became heavily dependent on the sea, and particularly the whaling trade. As such, Nantucketers did their best to remain neutral before and during the war; however, these actions proved damaging and resulted in the prospect of attack from both sides (Spence 1980:57). Not only were the British actively engaged in crushing the colonial whale fishery, but early in the conflict, American colonial leaders obtained reports that supplies were being transshipped to the British via Nantucket, and therefore, sanctions were laid against the islanders (Taylor 1977:583).

In all colonial business ventures, the war had a devastating financial effect. For whaling, almost the entire colonial fleet was destroyed. Nantucket was particularly hard hit: of its 152 vessels operating at the outbreak of the war, 124 were captured by the British and 15 others were wrecked (Fairburn 1945:990). Those employed in many other branches of the industry were also affected; rope makers, coopers, blacksmiths, carpenters, and shipwrights all saw their businesses diminish (Macy 1835:68). Thus, in communities that shared ownership of vessels and voyages, the war presented twice the problems. If the vessels were not able to engage in the fishery, then no money could be made; and if the vessels already at sea were captured, then their investment was lost. Further, since British ports were closed to American vessels, what oil the Nantucket merchants had was not able to be traded directly, and thus, that revenue was lost as well.

As a general policy throughout the war, British privateers and naval ships aggressively pursued merchant vessels and interrupted commerce to starve the colonies into submission. And since much of the revenue of the colonies came from the proceeds of whaling, the British were determined to crush it. Still struggling to establish a profitable southern whaling industry of its own, however, England took full advantage of the situation and seized American whalers. When whaleships were captured, the vessels were either burned and sunk or taken as prizes, and the crews were generally given the choice of being imprisoned, being forced to join the Royal Navy and fight against their colony, or being impressed to serve on British whaleships. Many American whalers opted for the latter, and in doing so began to transfer knowledge of the methods they had developed. Though the government's efforts to stimulate Great Britain's whale fishery through the bounty system was not as successful as they had hoped, the Revolutionary War and ensuing conditions in America produced a British whaling business operated by New England mariners (Fairburn 1945:1008).

The constant harassment at sea, mistrust on the home front, and the loss of vessels by capture and forced donation to the war effort led New England whalers to cease operations for several years (Francis 1991:63). As the war dragged on, the situation in Nantucket became desperate. With no other way of obtaining the revenue needed to get supplies for the island's inhabitants, Nantucket whalers petitioned both sides of the conflict for consent to undertake limited whaling operations unmolested by naval or privateering actions (Spence 1980:57). In 1781, British authorities issued 24 permits that allowed named Nantucket vessels to undertake whaling without fear of capture. Over the following two years, 35 out of a total of 41 voyages ended in success with the others captured or burned for defiance of their papers (Hawes 1924:86). In 1783, the American Congress finally granted similar permits to 35 vessels, but this did little good, as by that time the war was ending (Jenkins 1921:231). While these permits did help to sustain Nantucket during the final years of the war, the industry was in ruins.

Post-Revolution Depression, 1783–1815

While the Treaty of Paris brought an end to the American Revolution in 1783, it also brought hard times to the North American economy (Woodward 1968:142). Determined to rebuild their industry and return to prosperity, whaling merchants in the ports of Nantucket and New Bedford quickly rounded up and repaired decaying hulks, as well as building new ships (*North American Review* 1834:101). As soon as they could continue their trade safely, whalers outfitted voyages from these ports. Owing to the lapse in hunting during the war, the whales proved to be easy prey, and soon, other New England ports followed suit (Tower 1907:40). Although whale products commanded good prices for a time, this boom was short-lived.

To protect British whaling interests, Parliament effectively closed the London market for American oil by imposing an £18 per ton duty in 1784. This crippled American efforts to revive the industry by taking almost all profit out of the equation (Whipple 1979:57). To aid the fishery, the Massachusetts legislature attempted to stimulate recovery by offering a bounty on whale products returned by vessels owned and operated by state residents. This attempt proved ineffective, however, because the demand for whale products had declined (Davis et al. 1997:37; Starbuck 1878:78–79). The lack of available whale oil during the war had led to the necessary substitution and acceptance of lower-quality tallow candles, which meant oil resulting from the bounty could not be absorbed by the young American market and overproduction lowered prices (Jenkins 1921:231).

Around this time, the British government's aspiration to make their whaling industries completely self-sufficient led to a shift in focus from the northern to the southern fishery with support for the trade coming from the highest levels (Chatwin 1996:25). Thus, spurred on by the protective tariffs and generous subsidies from the British government, London whaling merchants further expanded their operations into the South Atlantic (Richards 1994:19). The Southern Whale Fishery—so called to distinguish it from the Northern (Arctic) Whale Fishery—proved successful due largely to both the knowledge of American whalers impressed into service during the war and the enticement of many of those same whalers to continue working for the British thereafter (Spence 1980:57). The skills of Nantucketers were highly regarded and in great demand, so much so that some firms insisted on employing Nantucket whalers on each ship (Fairburn 1945:989).

During this time, the British also mounted a campaign to encourage Nantucketers to move their operations and to join Great Britain's newly resurgent whaling industry (Whipple 1979:57). Due to the stagnant conditions of the fishery in America, some Nantucket whalers migrated with their families to a new British settlement in Nova Scotia and others looked directly to the UK. Of note among these expatriates was William Rotch Jr., a shrewd Nantucket whaling merchant who represented a group interested in settling in England. Rotch approached the British government about establishing a whaling colony there on the condition that his group would be reimbursed for all expenses and that they would be allowed to collect the bounty that the British whaleships received. When the British government refused to grant these demands, Rotch, knowing that the French had long been interested in Nantucket oil, then petitioned Louis XVI's government with a similar deal (Bockstoce 1984:529). Interested in reviving French whaling to compete with the British, the French quickly agreed and provided handsome bounties. In 1784, Rotch and a small group of families established a whaling colony at Dunkirk, in northern France (Richards 1994:29).

A direct result of the establishment of Dunkirk was the gradual return of Nantucket as the leading whaling port in America (Stackpole 1953:134). Opening the French market to American whalers in 1789 infused the Nantucket industry with new life (Tower 1907:43). Ships sailing from Dunkirk soon competed with British ships on the Brazil Banks and in the South Atlantic grounds. Dominated by Rotch-owned vessels, the Dunkirk fleet expanded from 6 ships in 1786 to 26 in 1792, and whalers from New England continued to send their products to French ports (Dolin 2007:179). It also was during this dynamic period that the French Revolution began, and, as a result, France declared war on the UK in early 1793. At the outbreak of this conflict, the safety of American inhabitants in the Dunkirk colony came into question, and they opted to return to America. Among those who left was Rotch, who then settled in New Bedford—partly because he realized its potential as a mainland whaling center and partly because he found himself unwelcome among Nantucketers who considered him a deserter (Ashley 1926:32). This choice would prove very wise, as in a relatively short time, New Bedford eclipsed Nantucket and took the reins as the premier American whaling port.

During this period, the British Empire sought to make up for its loss of the American colonies and outfitted ships for voyages of exploration into little known regions. As far as British whaling industries were concerned, the most important of these expeditions were those of the famed Royal Navy navigator Captain James Cook. Not only did he identify previously undocumented lands, Cook also made observations regarding the natural world on three major voyages of exploration from 1768 to 1779. His identification of new territories such as Australia, New Zealand, Tahiti, and the Hawaiian archipelago proved to be very lucrative for the British in the coming decades and had a profound impact on the development and operations of both the British and American whale fisheries (Freeman 1951:74–81).

In the meantime, an increasing demand for oil had a disastrous consequence on whale populations in the known hunting grounds. By the late 1780s, both American and British whaleships hunting in the South Atlantic found that the decline in whale populations meant that longer voyages throughout the region were necessary to obtain cargos (Whipple 1979:57). Knowing that merchants and sealers had already undertaken voyages into the Pacific Ocean and reported an abundance of whales there, the idea of making the dangerous passage around the tip of South America soon became a reality. In January 1789, the whaleship *Emilia*, owned by the London whaling firm Enderby and Sons, rounded Cape Horn and found numerous whales off the southwestern coast of Chile (Spence 1980:63). Captain James Shields, a Nantucketer, wasted no time in chasing them, and soon, first mate Archelus Hammond, also of Nantucket, harpooned and killed the first sperm whale known to be taken in the Pacific (Stackpole 1953:145; Starbuck 1878:90). More whales were quickly taken, and, once full, the ship returned to London, where news of a rich sperm whaling ground spread like wildfire. Immediately, *Emilia* and nine other London whalers were dispatched for the Pacific (Richards 1994:26; Stackpole 1972:129).

By 1791, reports of the successful voyage around Cape Horn reached New England and vessels from Nantucket first went into the Pacific Ocean (Macy 1835:141). Nantucket, however, was not the only port outfitting vessels bound for the new grounds. The New Bedford whaler *Beaver* claimed the title of the first American whaleship to round the Horn; between the years 1791 to 1793 its captain noted that, of 40 whaleships sighted during that time, 10 hailed from New England (Dolin 2007:182). At first the whalers only needed to stick to the southern coast of Chile to quickly find their prey, and soon, this area became known as the On-Shore ground. The abundance of sperm whales found there meant that vessels could fill their holds and return to port in less than two years (Rydell 1952:62). As more and more ships came, however, they soon found it necessary to extend the scope of operations, and by the end of the century, whaleships cruised the entire western coast of South America (Hohman 1928:37).

Though the outlook seemed good for American whalers, the outbreak of the French Revolution created difficulties by destroying the burgeoning oil market. The captains of whaleships in the Pacific had no way of knowing that the bottom had dropped out of the market. Thus, they continued whaling at a feverish pace, and the oversupply of whale products meant that prices dropped still further (Rydell 1952:62). The glut in the whale oil market in turn created hostilities between the United States and France, and by 1798, the prospect of war between the two nations led French privateers to prey upon American commerce (Jenkins 1921:232). The whaling industry suffered heavily during this "quasi-war" with France (Stackpole 1972:285). Not only were several vessels captured, but many merchants refused to send their ships out due to both the threat of loss and the increased cost of insurance (Tower 1907:43). The political and military chaos in Europe placed all merchant vessels in a precarious position, as British privateers also roamed the seas for the next decade raiding commercial vessels and impressing sailors into service (Hohman 1928:37).

As a result of these and other privations, diplomatic relations between the United States and Britain gradually worsened over the first decade of the nineteenth century. In response to French and British harassment, President Thomas Jefferson signed the Embargo Act of 1807 and closed US trade with foreign countries. Though well-intentioned, the economic problems it created had a severely negative effect on US commerce (Woodward 1968:143). As the act resulted in little if any impacts on either the French or British economies, it was soon repealed and replaced with the equally ineffective Non-Intercourse Act of 1809, which opened trade to all countries of the world excluding France, Britain, and their dependencies (Coles 1965:11). Over the next three years, relations between the United States and Britain further deteriorated, and ultimately, the two countries declared war.

As with the American Revolution, the War of 1812 had a devastating influence on the American whale fishery. Ostensibly fought to defend the rights of American commerce on the high seas (Woodward 1968:143), both Royal Navy activities and lack of access to markets drastically impacted whaling during the war. Believing that a peaceful resolution to the causes of contention between the two countries would be found, many New England whaleship owners fitted out ships prior to 1812. Thus, when war was declared, many vessels hunting in both the Atlantic and the Pacific deemed it prudent to set a course homeward in the hope of avoiding interception by the enemy (Clark 1887:141). Captured ships were burned and their crews impressed into British service, and again, it was Nantucket that suffered the most—"in 1812, islanders owned and operated 116 whaling vessels, but by 1815 only 23 of them remained" (Hohman 1928:39).

Still needing to supply the market in London with oil, British whalers during this time continued fitting out trips to the Pacific Ocean. To ensure dominance by eliminating competition, the ships of the South Sea fleet were armed "so that each one constituted, in itself, a letter of marque as well as a whaleship" (Stackpole 1953:262). It did not take long for the US government to realize the defenseless condition of American commerce in the Pacific; to guard those interests they dispatched the two-gun frigate USS *Essex*. Under the command of Captain David Porter, the warship not only protected American merchant activities but also effectively destroyed the British whale fishery in the Pacific (Clark 1887:143). Porter's general strategy involved the element of surprise; a common ruse involved flying a false Union Jack when approaching enemy vessels, then running up the Stars and Stripes and easily seizing the lightly armed and completely surprised whaleships (Norton 2012:7). Being far from home and ports that could be guaranteed to remain neutral, he did not sink the ships he captured—instead he claimed them as prizes and used them to build up a small fleet of support vessels (Thurn 1915:314). Replacing the loss of these ships was not only expensive and difficult it was also considered too risky due to the presence of an effective US naval force in the Pacific (Chatwin 1996:27). Though USS *Essex* was captured by the British off the coast of Peru in 1814, Porter and his crew succeeded in taking 12 of the 20 or so British whaling ships operating in the Pacific during that time (Jones 1981a:22). Thus, America not only achieved its naval objectives but also greatly helped to turn the tide of superiority in the Pacific sperm whaling industry.

The Golden Age of American Whaling

By the time the Treaty of Ghent officially brought an indecisive end to the war in December 1814 (Coles 1965:237), the New England whale fishery was effectively in ruins. As soon as news of the treaty's signing reached the United

States, however, whaling firms began fitting out vessels for voyages. Nantucket and New Bedford whaling merchants gambled on the idea that demand for oil would bring good prices. They wasted no time in outfitting whatever vessels they could find and sending them to the Pacific, where from prewar experience they knew they could expect vast returns in the shortest amount of time. As it turned out, their gamble paid off handsomely (Rydell 1952:64).

William Fairburn (1945:1001) points out that American whaling "dragged" in 1816–1817 because it took time for the ships that had sailed for the Pacific in 1815 to return. This time was not wasted, however, as they built new vessels, secured supplies, and received incoming ships hunting the closer Atlantic grounds. As soon as the first of the postwar whaleships entered their home port, the wharves were again stacked with greasy casks, and the industry entered a period of unmatched prosperity (Ellis 1991:159). Through the resilience of the industry and the productivity of the Quakers, the United States quickly took the lead over other nations engaged in Pacific whaling. Thus began the era that American whaling historians have deemed the "Golden Era of Whaling" (Tower 1907) or "Golden Age" (Davis et al. 1988; Davis et al. 1997).

As the numbers of ships hunting in the Pacific increased, an expanding knowledge of its havens and hazards developed. Though it is true that European explorers such as Cook had canvassed parts of the Pacific long before American sealers, traders, and whalers, none accomplished as much for understanding its geography and oceanography (Fischer 2002:95). Throughout the Pacific Ocean are thousands of islands, reefs, and shoals that were sighted and named by European whalers; when an island or group of islands was encountered and found to have friendly inhabitants offering fresh food and clean water, the location(s) were noted in logbooks and journals and eventually made it onto nautical charts (Boggs 1938:185; Reynolds 1835:712). Where possible, information pertaining to sea conditions was noted, since such data could later help to determine the best approach to an island and warn of navigational hazards. Best efforts were also made to obtain precise locations of the dangers they encountered, such as partially submerged reefs and shallow shoals, so that others could avoid the potential for shipwreck. Many of the islands and atolls still maintain these names, some after the ship and some after the master that encountered them (Leff 1940:5). Thus, the names of Howland Island, Pearl and Hermes Atoll, Maro Reef, and Gardner Pinnacles are among many that are testament to the American and British whalers' impacts on the identification of the geography of the Pacific.

In the decades of the early to mid-nineteenth century, the number of ships engaged in American whaling grew exponentially. From the few ships sent out at the end of the War of 1812, the number of vessels grew to 400 by the 1830s and over 700 by the late 1840s (Rydell 1952:67). This expansion in turn resulted in rapid exploration of the Pacific, as ships searched for both new hunting grounds and for islands where wood and water could be obtained easily (Thomson 1914:12). In their exhaustive analysis of the economic history of American whaling, economic historian Lance E. Davis and colleagues (1997:38) determined that in addition to the increased number of whaleships and whalers employed in the Pacific region, the industry's growth during this period had three other major dimensions: the size and rigs of ships employed, the number of ports engaged, and the more and ever distant grounds they hunted. Each of these interrelated factors contributed not only to the success of American whaling in this period but also to the impacts it had on the Pacific region.

The postwar revitalization of American whaling signaled the need for larger ships. The three main factors affecting choice and layout of vessels employed at this time included the length of voyages brought on by the focus on ever distant grounds, the need for self-sufficiency due to the distance from safe harbors for reprovisioning, and the potential for higher financial returns on increased cargo sizes due to imagined unlimited stocks of whales. Thus, as the industry grew, reliance on the small sloops, schooners, and brigs used in the eighteenth century yielded to much larger and stauncher vessels (Davis et al. 1997:38). Though in the early 1820s, brigs of up to 200 tons were still employed, those were soon replaced by ships and barks of greater tonnage. In fact, from then until the 1850s, the tonnage of whaleships increased fivefold (Fairburn 1945:1627) and, at its peak, accounted for almost 7% of all registered merchant shipping tonnage in the United States (Chatwin 1996:6). Over the course of the golden age, the size of vessels increased to an average of around 400 tons.

Not all vessels were newly constructed. Many whaling merchants sought only to achieve the highest return on a voyage and rushed to get as many vessels to sea as possible. As a result, older vessels used in other industries were commonly purchased and refit for whaling. The most important deciding factor for whaleships in this period was cargo capacity. As their purpose was simply to fill their holds with oil, they had no need to achieve the speed of clippers. Instead, tub-like vessels with bluff bows were generally preferred as they provided stable working platforms.

Whaling firms understood that regardless of whether old or new, their vessels would likely return from the hunt filthy and scarred from encounters with whales, reefs, and shoals—despite the best efforts to keep them clean. Historian A. Hyatt Verrill (1916:56) describes the relationship between the captains and their vessels by stating, "As long as the ship held together and was able to weather the seas and gales, as long as it would carry its cargo of oil and bone, as long as the patched and dingy sails would serve to catch the winds and carry the whalers hither and thither, the whalemen were satisfied, and some of the old hulks, which were used for whaling, would appear fit only for the scrap pile to a merchant sailor." Thus, the dingy appearance of their smoke-blackened sails, the large brick tryworks on their decks, the cutting tackle attached to their rigging, and the multiple boats hanging conspicuously from wooden davits made the typical whaleship of this period unmistakable (Hohman 1928:40).

The increase in vessel size needed for extended voyaging eventually had a direct consequence on Nantucket's significance to the industry. Since the early 1700s, it had been the dominant whaling port in New England, and its postwar success had a stimulating effect on the region (Tower 1907:49). The tradition of young Nantucket whalers rising through shipboard ranks with the goal of becoming captains meant that they were extremely skilled in all aspects of the operation and that they had an ingrained sense of pride in their profession and duty. Despite all their knowledge of the business, skills as seafarers, or sheer determination to constantly rebuild, however, the physical limitations of Nantucket's harbor ultimately became too great. As the size of the ships grew larger, the shifting sands of the harbor bar at the entrance to Nantucket Sound became too dangerous, and its shallowness greatly restricted the passage of ships with full cargos. In 1839, a steam-powered floating dock was built to assist ships in passing, but it was too late, and by 1869, whaling from the island ended (Ashley 1926:33).

In the meantime, the success enjoyed by early postwar voyages sent out from Nantucket and New Bedford led many expatriate whalers to return to the United States to take up the trade (Spence 1980:83). Merchants in many northeastern ports quickly realized the economic potential of the industry, and soon, harbor towns in Connecticut, Rhode Island, New York, and other parts of Massachusetts engaged in whaling. The rate at which New England ports joined in the rush to the Pacific can be seen as an indicator of the promise of whaling at this time; in 1820 there were 16 such towns, by 1835, there were 32, and by 1841, 38 New England coastal towns were fitting out whaling ships mostly bound for the Pacific (Davis et al. 1997:41).

Of all the New England ports, none achieved as much success as New Bedford. Established in 1765 by successful whaleship owner and Quaker Joseph Rotch (Hegarty 1960:9), the port developed slowly and quietly until it eclipsed Nantucket to become the nation's whaling capital. Though Nantucket's location farther out to sea and closer to the Gulf Stream and Atlantic grounds was advantageous in the eighteenth century, the opening of the Pacific brought with it strategic changes. Located inshore of Nantucket on the Massachusetts coast, New Bedford offered a spacious deepwater harbor and eventually provided direct access to a railroad network that connected major cities in the United States. Adding to the better physical location was the dogged determinism of the Quaker shipowners, described as "tight-fisted, cruel, ruthless a set of exploiters as can be found in American history" (Morrison 1921:315). These "oil barons" quickly established refineries, cooper's shops, tool shops, and many other subsidiary industries to become the leaders in both whaling craft production and the nation's fifth-largest port for shipping (Fairburn 1945:1002–1003; Morrison 1921:315). By the 1850s, New Bedford's importance to whaling grew to the extent that almost half of American whaleships called it home (Davis et al. 1997:41).

Both the greater number and size of ships and the ambitions of whaling interests in towns throughout southern New England led to a constant search for rich, new whaling grounds. The high demand for sperm oil and spermaceti, coupled with the supposition that Pacific sperm whale populations were limitless, created a rush to the region. Even as the century progressed and other illuminants came into use, the considerably viscous quality of sperm oil saw its main use shift to lubrication of light, rapidly moving machines (Davis et al. 1987:5). Therefore, demand for sperm whale products during the golden age meant that profits from voyages could be unbelievably high. Despite the potential dangers in reaching them, the richer the grounds, the easier the hunting, and the more quickly holds could be filled. These factors in turn yielded even higher returns, as the cost of outfitting was reduced by the shorter voyages.

The increasing number of ships operating in the region soon diminished the stocks of whales on the On-Shore grounds along the coasts of Chile and Peru. While expansion beyond those waters simply followed the century-old pattern of exploiting one ground and then moving to another, the American fleet moved on with surprising speed (Kugler 1971:24). Unlike previous Europeans who explored the waters of the Pacific, American skippers were pragmatic businesspeople who favored financial rewards over glory (Fairburn 1945:512).

The first of the new hunting areas to be located was the Off-Shore grounds. In 1818, the Nantucket whaleship *Globe* sailed out from the On-Shore grounds and their proximity to the South American coast to find this new area, which is located south of the equator between 105° and 125° west (Bockstoce 1984:531). The discovery of this rich ground was followed quickly by the location of the Japan Grounds in 1820 by the whaleships *Maro* of Nantucket and *Syren* of London. With sperm whale populations abounding in the summer months, this ground stretched from Midway Island west to the Bonins (Kugler 1971:24). The same year saw opening of the rich On-the-Line grounds along the equator and among the Gilbert Islands, as well as the Middle Ground in waters between New Zealand and Australia and the Vasquez

Grounds to the north of New Zealand toward Tonga (Ward 1960:44). The 1820s also saw the rise of country whaling, in which ships cruised relatively shallower water among reefs and islands in the central Pacific in search of prey (Rydell 1952:67).

During this time, American whalers also operated in minor grounds throughout the Pacific such as those around Micronesia. They also hunted in the Indian Ocean, among the islands of Indonesia and the Philippines, though none of these were ever as important to the sperm whale fishery as the aforementioned major Pacific grounds (Johnson 1995:49). Of all the whaling grounds explored and exploited in the first half of the nineteenth century, probably the most productive were the Japan Grounds (Richards 1999:190). By the 1830s, two patterns for cruises were established depending on the season: (1) whalers sailed for the Pacific via the Cape of Good Hope, across the Indian Ocean to Timor, then north between the various island chains, or (2) they entered by way of Cape Horn to the Off-Shore ground, then stopped at Hawaii to rest and reprovision before sailing for the Japan Grounds (Kugler 1971:24). The development and increasing importance of ports in the Hawaiian Islands is discussed in more detail in chapter 3.

The increase in the numbers of ships hunting the Japan Grounds eventually brought more discoveries in the North Pacific, including grounds toward Alaska and the Aleutians, which were identified around 1835 (Johnson 1995:48). The main species of whale found on those grounds was the North Pacific right whale (*Eubalaena japonica*) (Richards 1999190). Grounds in the Arctic, such as the Northwest Coast and the Kodiak, became more significant, as by the early 1840s sperm whale populations were noticeably declining (Bockstoce 1984:531). As Davis and colleagues (1987:5) explain, the quantity of sperm oil that the fleet harvested steadily increased through the 1830s, but by the late 1840s, the catch declined until the end of the century.

Declining sperm whale populations meant ever longer voyages to obtain a profitable cargo. Extended cruises in turn increased the costs of outfitting and labor, as well as raised the risk of disaster. The combination of these factors, the need for affordable lighting fuel, and the immense whale populations on the Arctic grounds soon led more vessels to hunt and explore the region. Thus, a pattern developed whereby sperm oil was the object of the hunt while ships cruised in the south and central Pacific, but when they entered the northern waters, their attention shifted to bowheads (*Eubalaena mysticetus*) (Starbuck 1878:771). For a time, this pattern provided the markets with both types of oil, and the industry maintained a healthy return on voyages.

Another market factor that drew ships into northern waters with greater frequency was increasing demand for baleen of right, bowhead, and humpback whales. Generally referred to as "whalebone," baleen plates are situated in a whale's mouth (Lawrence and Staniforth 1998:5) and were a highly valued industrial commodity. Historian John Bockstoce (1984:531) describes the versatility of baleen by stating that it "could be cut into long strips or indeed any shape that the baleen plate would allow, without sacrificing its strength or springy flexibility, and could also be moulded by steam to hold a new shape, it was used in a number of items where resilience was required, among them umbrella spokes, buggy whips, hat brim stiffeners, brush bristles, and corset stays." As a result, the demand for whalebone skyrocketed, and whaling merchants began to realize higher prices for cargos of whalebone and whale oil than for sperm oil. An industrial boom occurred with the opening of the Kamchatka and Sea of Okhotsk grounds in 1846–1847. By the early 1850s, the Arctic bowhead whale fishery was underway, and by the 1860s, almost the entire Pacific fleet cruised the area (Jenkins 1921:236). Success in the Arctic fishery meant hunting among ice floes, an activity that came with increased risks. Preventative measures were needed, such as reinforcing vessels' hulls by placing additional strengthening timbers in their bows and adding extra layers of planking to avoid puncture by ice. It only took a little over half a century for whalers to deplete the sperm whale populations of the Pacific and to penetrate the Arctic, which in the coming decades became the most important of all the whaling regions (Tower 1907:59–60).

The Decline of American Whaling

The apogee of American whaling occurred from 1852 to 1858 with the industry maintaining its profitability by focusing on the Arctic grounds (Fairburn 1945:1038–1039; Hohman 1928). By the end of the decade, however, several factors took a toll on operations, which in turn led to a rapid decline in tonnage of the whaling fleet throughout the 1860s (Table 2.1). Some reasons for this decline were directly attributable to the practices of whaling firms. These included overexploitation of whale populations, increased costs of outfitting, less profitable returns on whaling ventures, necessarily longer voyages of the ships, and greater difficulty in obtaining competent crews (Fairburn 1945:1039). In addition, some uncontrollable factors external to the industry also had devastating effects.

The first of these external factors was the mid-nineteenth-century California gold rush. The discovery of gold at Sutter's Mill in 1848, and the "get rich quick" stories that followed, led many people to leave the East Coast of the United States for the West. The main port of call for these would-be miners was San Francisco, which by this time was also a resupply and transshipment point for whaleships. As news of the gold rush spread among whalers, many saw it as an opportunity to give up a hard life at sea and soon deserted. As the price of passage to the West increased, those familiar with the sea knew that

Table 2.1. Decline in Total Tonnage of the American Whale Fleet for Selected Years, 1859–1867

Year	Registered Tonnage	Percentage Decline from Average Tonnage, 1852–1858
1859	185,728	3
1860	166,841	13
1861	145,734	24
1862	117,714	39
1863	99,228	48
1864	95,145	50
1865	84,233	56
1867	52,384	73

Source: From W. A. Fairburn (1945).

they could obtain a free trip out by signing on to a Pacific whaling voyage and then deserting the ship when it arrived in San Francisco. This practice became so prevalent that sometimes whole crews, often including the captains, deserted en masse (Robotti 1962:196).

The gold rush negatively affected the industry not only by luring away experienced whalers but also by reducing the number of vessels employed in the industry. This was due to both their reconfiguration for use as passenger vessels and to the abandonment of large numbers of them along the northern California coast. In turn, this greatly reduced the numbers of New England ports fitting out voyages. By the late 1850s, only New Bedford and its neighboring towns, such as Mattapoisett, Fairhaven, and Dartmouth, maintained an increase in vessel numbers and industrial investment (Hawes 1924:196).

Another turning point for the fishery was the US Civil War (1861–1865) (Kugler 1971:26). As with the Revolutionary War and the War of 1812, the Civil War had a crippling effect on American whaling (Dolin 2007:309) through economic sanctions, a lack of demand for oil resulting from decreased capital, and targeted destruction of the fleet. While the legality of slavery can be considered the main issue that led to the war between the Union and the Confederate States of America (CSA), there were many other economic and political differences between the states that supported its outbreak. As such, the impacts of the war on commerce and trade were a by-product of these larger national issues. Of the wartime actions that affected the output of whaling during this time, perhaps the two most significant were those that contributed to the size of the fleet.

The first of these factors was the CSA's strategy of disrupting commerce through attacks on whaleships. This tactic was executed by the Confederate raiders CSS *Alabama*, operating in the Atlantic Ocean, and the CSS *Shenandoah*, operating in the Pacific. Together, the actions of these two vessels resulted in the loss of 80 whaleships, several of which were attacked after the war officially ended (Kugler 1971:26; Starbuck 1878:103). The other major military strategy that affected the whaling fleet was the Union's attempt to blockade harbors in the southern ports of Charleston, South Carolina, and Savannah, Georgia. Known as the Stone Fleets, US naval officials purchased some 40 whaleships from various New England ports, filled their cargo holds with stone, sailed them south, and sunk them in strategic positions to foil blockade runners and privateers (Starbuck 1878:101). Problems encountered with positioning the vessels during their scuttling were compounded by tropical storms and the "the ceaseless flow of the sea," which quickly created channels between the ships, thus making the submerged walls largely ineffective (Songini 2007:140).

Though after the war the prices for whalebone rose dramatically, the return was insufficient for most firms to justify outfitting many ships (Heffer 2002:53). As both the grounds and the types of whales hunted changed, the need for newer technologies became apparent when considering the cost of replacing vessels lost during the war. Thus, over the next few decades, the number of sailing ships hunting whales dwindled and more steam-powered vessels were brought into service. Overall, the Civil War is estimated to have reduced the number of ships in the whaling industry by 60% between 1860 and 1866 (Morgan 1948:143; Vance 2002:41).

The most portentous factor contributing to the eventual demise of American whaling was the introduction of alternatives to whale oil. As supplies of whale oil decreased and became little more than lubrication for industrial machinery, the average American household turned to burning tallow candles and lard as illuminants. By the late 1820s, products such as a coal-gas by-product called benzene, "rosin oil" or "burning fluid" made from pitch pine, and a turpentine distillate known as "camphene" had all been tested as possible alternatives to whale oil (Beaton 1955:30). Though these fuels were consumed to some extent, their volatility, smell, and the smoke produced generally precluded their everyday use, so whale oil remained the fuel of choice. This situation, however, soon changed with the discovery of petroleum in 1859.

First discovered in the United States in Titusville, Pennsylvania, the value of petroleum as a substitute for whale oil was realized almost immediately (Daum 1959:21). Though in the years prior to its discovery, experiments with obtaining oil from coal in England were successful, it was a distillation process for manufacturing kerosene patented by A. P. Gesner that brought about a revolution in lighting and lubrication (Gesner 1861). As the refining processes for kerosene improved, it quickly began to rival other oils (Tower 1907:77). Seemingly overnight the groundwork for mass consumption of petroleum oils was laid, and by the fall of 1859, at least 33 "coal-oil" refineries were actively engaged in the manufacture of hydrocarbon illuminating oils (Beaton 1955:28). The by-products of the refining processes were also found to be excellent lubricants, which further added to the versatility of petroleum. Ease of extraction, simplicity of refining, ever increasing discoveries of deposits around the county, and growth in its use as machinery fuel are among the factors that allowed the industry to quickly dominate.

The development of the petroleum industry had devastating consequences on American whaling. As it also coincided with the Civil War, by the mid-1860s the more expensive whale oils had largely been abandoned for cheaper petroleum products. As petroleum use grew, even the whaling merchants of New Bedford realized that it would be the end of their industry, and therefore, they invested in construction of a petroleum refinery (Ellis 1991:166). Thereafter, demand for baleen kept the whaling industry afloat, and the ever diminishing fleet concentrated their hunting efforts on the Pacific Arctic (Kugler 1971:26). The focus on the Arctic region and its fast-swimming bowhead populations meant that use of steam-powered vessels and bow-mounted harpoon guns became the rule. Thus, by the last quarter of the nineteenth century, the sperm whale fishery, and indeed the old way of whaling, was all but gone.

Conclusion

The evolution of American whaling from its European origins led to organizational and technological adaptations in the second half of the eighteenth century that enabled it to become a highly successful, global enterprise. Despite poor diplomatic relations with the UK in the late eighteenth and early nineteenth centuries, which hampered its operations, the American whale fishery flourished when the fleet began to hunt throughout the expanse of the Pacific. American whalers found great success on Pacific whaling grounds, which resulted in rapid expansion of the fleet and in US prominence in the region from roughly 1820 to 1870. Several factors, mainly uncontrollable, however, led to industrial decline, and in the decades that followed the golden age, American whaling became obsolete.

3

Hawaii as an Entrepôt for Pacific Whaling, 1819–1870

The boom that occurred in whaling after the War of 1812 had a profound impact on the trajectory of the American whale fishery. The opening of rich hunting grounds, coupled with the establishment of regular trade in the Pacific Ocean at the end of the eighteenth century, not only increased geographic and economic expansion, it also resulted in a fundamental change to the way American whaling agents operated. To maximize efficiency, agents increasingly used the practices of resupplying vessels at established locations and transshipping oil cargos. This change in business strategy had a dramatic effect on the success of the American fishery, particularly through the transformation of quiet harbors in the main Hawaiian Islands into bustling centers of trade and activity.

Colonization and European Trade

The Hawaiian archipelago, located in the central North Pacific Ocean, comprises a grouping of larger islands, commonly referred to as the "main Hawaiian Islands" at its southern end, and a chain of islands, atolls, and shoals known as the Northwestern Hawaiian Islands, which extend to the northwest (PMNM 2011:1). The geographical location of the islands provides a subtropical climate that offers abundant freshwater sources and a highly productive agricultural environment. It is these characteristics that have attracted people to the islands for nearly two thousand years.

Polynesian voyagers are considered the first people to have inhabited the Hawaiian Islands. They arrived in the latter phase of a two-pronged Pacific colonization movement that archaeologists and linguists generally agree began approximately 3,500 years ago (Bellwood 1979:282; Russell 2009:63). Though the exact arrival date is unknown, research suggests that double-hulled voyaging canoes from the South Pacific most likely encountered Hawaii around AD 1000 (Kirch 2011:22; Van Tilburg 2002a:248). This contact established communication between two distant areas of the Pacific and contributed to the development of complex cultural traits that manifested socially, linguistically, and technologically (Van Tilburg 2002a:248).

The appearance of Europeans approximately 500 years ago marked the start of the great age of European exploration in the Pacific (Daws 1968:xi). Beginning with sixteenth-century Spanish navigators searching for an alternate route to trade with the East Indies, European exploration and contact had dramatic effects on the economic and cultural development of the Pacific islands. Thus, the Hawaiian Islands escaped Spain's attention and remained isolated and unknown to Europeans for over 250 years (Daws 1968:xii; Gschaedler 1948:302).

Spanish success in establishing a reliable trade route across the vast Pacific spurred other European countries to explore in the hope of expanding their colonial networks. Seventeenth- and eighteenth-century expeditions undertaken by Dutch, French, and British navigators resulted in significant geographic and scientific discoveries throughout the region. It was during one such expedition that Europeans first documented the Hawaiian Islands. On his third round-the-world voyage for the Royal Society of London, Captain James Cook sighted them in January 1778 (Coulter 1964:256). In honor of the first Lord of the British Admiralty—and an important backer of the expedition—Cook called the group the Sandwich Islands and that name was commonly used until the 1840s (Clement 1980:55; Manchester 1951:80). Bound by orders from the British Admiralty, only brief landings were made to secure provisions at the islands of Niihau and Kauai. Soon thereafter, Cook headed for the northwest coast of North America for mapping operations; upon completion of that task, he returned to Hawaii to explore the main group of islands (Manchester 1951:80). Though no problems with Native Hawaiians were experienced in the subsequent months, just prior to departure for England, in February 1779, Cook was killed in a dispute with some of the chiefs at Kealakekua Bay (Gilbert 1926).

In the years following the return of Cook's ships to England, published accounts of the expedition contained detailed information about new geographical discoveries, as well as remarks pertaining to the commercial potential for establishing a trade in sea otter skins (Dodge 1976:55). First introduced to the sea otter through trade with Russians, the Chinese prized the fine, soft, thick fur (Dodge 1965:58). Acting on assertions made by Cook expedition chroniclers, by 1785 entrepreneurial merchants, first from the UK and then the United States, began to speculate on newly opened markets in Asia. They sent cargos of the highly desirable sea otter furs, as well as prized ginseng root, to China to be traded for equally valuable cargos of tea, nankeens, chinaware, silks, and drugs. Those early voyages proved successful and did more than just generate great profits for their investors; they laid the foundations for what would become known as the Old China Trade (Rydell 1952:23).

Though the British took the lead in the first years of the China Trade (Hezel 1983:133), pioneering Americans were soon exchanging goods with Native hunters of the Pacific Northwest and outpacing their competition. Though this practice was considered extremely dangerous at the time, the risks were abundantly rewarded with large cargos of inexpensive furs for which they exchanged goods such as "guns, ammunition, hardware, textiles, beads, and baubles" (Howe 1984:171; Miller 1988:8–9). The trade was given a major boon with the spread of news of an abundance of fur seals along the northwest coast of America.

By the late 1780s, American fur sealers developed a pattern for their voyages to best deal with seasonal sea conditions: they sailed from Canton (Guangzhou), China, to the northwest coast to hunt or trade for sea otter and fur seal skins; then replenished their supplies in Hawaii over winter, while avoiding dangerous storms; and then returned to China to sell the furs for a small fortune (Fischer 2002:95). The addition of seal skins to cargos destined for the China Trade caused dramatic increases in profit margins for Americans. Though European traders soon attempted to replicate this success, the persistent conflicts of the Napoleonic era depleted their capital and allowed merchant adventurers from the United States to ply the trade almost without competition (Kuykendall 1934:369).

By the early 1790s, the British monopolies in the Pacific were also engaged in the trade; the East India Company sold furs at the markets in Canton, while the South Sea Company brought them to the London market (Smith 2002:4). Though the fur seal trade remained profitable for the next 20 years, fierce competition between the two parties, compounded by almost constant warfare in Europe, ultimately resulted in the disappearance of English traders from it altogether (Howay 1932:7). Thus, American interests, mainly from Boston, dominated the trade in the first decade of the nineteenth century, and, despite fluctuations in prices resulting from oversupply and unpredictable international markets, they profited handsomely from it (Rydell 1952:28). Growth of this industry was, however, always to be limited by seal populations, and "the high prices obtained in the early years led to intensive exploitation and rapid depletion of known colonies" (Smith 2002:4).

Throughout its early development, the Hawaiian Islands proved to be an important component of the China Trade. The long passages between Canton and the northwest coast of America depleted provisions, and the time required on the return meant that crews suffered through violent winter storms in the North Pacific. A solution to both issues was soon found when traders realized that Hawaii was ideally equipped to operate as a provisioning point for ships bound for China and as a place to winter before returning to the northwest coast (Dodge 1976:17). The first American to contact the Hawaiian Islands was Captain Robert Gray, whose Boston trading vessel *Columbia* stopped there in 1789 in the course of his circumnavigation of the world (Tate 1980:1). Gray's reports of a "tropical" environment, safe harbors, and agreeable inhabitants soon resulted in the islands becoming a requisite stopover for most American traders.

In exchange for nails, hoop iron (for adzes), and cloth, the ships could obtain fresh water, fruits and vegetables, hogs, and salt, as well as useful goods such as rope made from native fibers (Fischer 2002:95). Other attractions included Native crafts, souvenirs that sailors took home to their families, alcohol, and partners with whom to spend the night (Dodge 1976:57–58). The incredible seafaring skills and fitness of Native Hawaiians was quickly recognized, and soon, they were recruited to fill positions of lost crew or deserters (Kuykendall 1934:369). Realizing the potential for obtaining greater power, by the 1790s local chiefs also wanted guns, powder, artillery, and small boats, which they used against their intra- and interisland rivals (Fischer 2002:95). Regardless of the exchanges and provisioning, the opportunity to easily break up the rough voyages in such a warm, subtropical location was seemingly worth any price, as the Hawaiian Islands were considered the "summer isles of Eden" (Howay 1932:10).

The lucrative fur trade began to decline by the end of the first decade of the nineteenth century due to the overexploitation of resources and financial crises (Smith 2002:4). As such, the strategy of traders changed, and the stopovers in Hawaii were all but discontinued. Instead, the ships spent the entire year moving between villages on the northwest coast trying to obtain suitable cargos (Howay 1932:10). But just as the fur trade was declining, the China Trade gained another boon through the introduction of sandalwood (*Santalum ellipticum*).

Europeans and Americans first discovered that sandalwood grew in the Hawaiian Islands around 1790. Capitalizing on the knowledge that the Chinese had a long history of using Asian sandalwood as incense and in the making of delicate furniture pieces (Tate 1980:1–2), a few American vessels began shipping it as a commodity to the Canton market. Although these early voyages were successful, they were discontinued after 1795 as King Kamehameha I consolidated power in Hawaii under a single monarchy (Rydell 1952:38). Once completed, trade in the islands was reinitiated, and in 1804, Hawaiian sandalwood was again shipped to China (Dodge 1976:61). Though considered inferior in quality to Asian varieties of sandalwood, centuries of careless exploitation had exhausted those supplies and in turn created a need for an alternative source (Oliver 1989:161). Thus, when sandalwood from Hawaii,

Fiji, and the Marquesas was introduced, it quickly became prized (Ralston 1978:13). By 1810, its use in topping up cargo holds of ships low on skins became standardized, and it was an important element of the China Trade (Dodge 1976:61).

Hawaiians and other Polynesians also valued the qualities of sandalwood and mixed it with coconut oil to be used as an aromatic body rub (Gibson and Whitehead 1993:159). Kamehameha I, however, understood its value as a commodity, and soon, Hawaiians were combing the archipelago for sandalwood (Manchester 1951:83). Seeing the scented wood as the key to developing an empire and establishing a Hawaiian navy, the king and his son, Liholiho (Kamehameha II), traded large quantities of the wood (Dodge 1976:61).

By 1823, however, the Hawaiian sandalwood boom was over. As with other commodities sold through the China Trade, overexploitation not only depleted resources it also resulted in oversupply in the Canton market. This glut drove prices down, and the Hawaiian suppliers never recovered (Ralston 1978:14). Though a small amount of the fragrant wood continued to be shipped from the islands by 1835 the resource was commercially exhausted (Freeman 1951:346). Though the loss of this trade might have resulted in decline of the then-dependent economy of the Hawaiian Islands, the discovery of nearby whaling grounds already supplemented the China Trade and brought with it an era of development that forever changed the islands.

Pacific Whaling and the Japan Grounds

Although the sandalwood era was brief and hectic, it had long-term significance in the Hawaiian Islands (Bradley 1943:278). Prior to sandalwood's development as a commodity, the islands served merely as a stopover point for vessels in transit; however, the increase in the number of ships resupplying or picking up cargos in the early decades of the nineteenth century resulted in increased importance of Hawaiian ports. As a consequence, Honolulu slowly developed into the "center from which radiated the trade routes to the eastern and central Pacific" (Bradley 1943:278). Its significance was further underscored when it became the monarchy's chosen hub and Kamehameha I took up residence in Waikiki in 1804 (Tate 1980:2).

Since the Kingdom of Hawaii had the only suitable ports within a 4,000-mile radius (Day 1966:169), American and British mercantile interests forged strong alliances with Hawaiian royalty and were soon allowed to establish agencies in Honolulu (Igler 2004:709). In the early period, shipping agents assigned to these houses performed a dual role: they "supervised the purchase of sandalwood and its shipment to Canton" and "were charged with the disposal, either in the Hawaiian Islands or along the American coast, of the cargos of American manufactures sent to the Pacific by their employers" (Bradley

1943:278). To provide the residents of the islands and visiting crews with easy access to increasingly desired goods, a few small retail stores were set up in Honolulu. As more merchant vessels visited, the harbor developed as a veritable entrepôt for European merchandise in the Pacific where consumers could find a wide variety of manufactured goods (Bradley 1943:278–79).

In 1819, both American and British whaleships first called at Honolulu, which was then nothing more than a cluster of grass huts and a few previously established storefronts (Allen 1973:181). While in port, the captain of a Massachusetts merchant ship reported having seen "great numbers of whales off the mysterious and forbidden Japanese Islands" (Fairburn 1945:512), and the whaleships that received this news immediately raced in that direction. Of the five that sailed, *Maro* of Nantucket and *Syren* of London were first to take whales, and in doing so are credited with opening the Japan Ground in 1820 (Gibbs 1977:59; Mrantz 1976:4). This expansive sperm whale haunt is recognized as being among the most productive ever identified, and though its boundaries were somewhat ambiguous, it essentially stretched from the top of the Northwestern Hawaiian chain into the open ocean toward Japan (Richards 1999:189).

Because Japan was closed to foreigners, this discovery proved to be a major catalyst in the development of Hawaiian ports as the principal provisioning points for whaleships working in the region (Kuykendall 1934:381). News of the productivity of the Japan Grounds spread like wildfire, and the number of ships calling at Hawaiian ports grew exponentially. Though in 1820 only 5 ships cruised on the grounds, by 1829 nearly 180 called at Honolulu while en route (Daws 1968:169; Mrantz 1976:9). Though the Japan Grounds were the main focus of the early years, new grounds opened in the North Pacific around 1835, followed by more in the Sea of Okhotsk in 1847, and finally those in the Bering Sea and Arctic Ocean just before 1850 (Johnson 1995:48). Each of these discoveries resulted in a better understanding of the seasonal movements of whales, which had a major impact on hunting strategies. These changes in turn contributed to the importance of the northern, rather than the southern, Pacific as the central focus of American whaling and made Hawaii an industrial crossroads (Johnson 1995:48).

Knowledge of sperm whale migration in the Pacific and the predictably unpleasant weather experienced during some parts of the year resulted in a reliable pattern for whalers and merchants alike. During the summer, ships hunted the fertile grounds in the North Pacific; as the water cooled and sperm whales moved closer to the equator and South Pacific, the whalers followed. Because a season of whaling generally exhausted the supplies of fresh water and provisions kept onboard, whaleships could easily visit Hawaiian ports to replenish while transiting in either direction. Thus, each spring and fall, many ships called in for two- to three-month-long layovers (Kuykendall 1934:381). Whaling captains realized that these stops provided not only the opportunity to resupply and refit vessels but also offered shelter from foul weather and gave sailors a welcomed break (Freidel 1943:381).

Adoption of this seasonal pattern by sperm whalers affected the development of the Hawaiian Islands. As the grounds were extended farther and the whale populations were reduced, the duration of the voyages became longer. Along with longer voyages, came other needs, such as secure ports equipped with dependable sources of fresh provisions and merchandise for replenishing ships' stores, facilities for repairing and refitting ships damaged while on the grounds, and recreation for the crew. To provide these necessities, each ship was required to pay a fee to the local harbormaster. Thus, the visits from whalers provided revenue that helped to develop port infrastructure (Gibbs 1977:60).

From the earliest visits, it was clear that Hawaii was to become an integral component of Pacific whaling and American commerce in general. Acting on this presumption, in 1820, President James Monroe stationed a commercial agent there to protect the interests of American merchants and seafarers engaged in the commerce of the North Pacific (Bradley 1943:279). The British government also appointed a consul to the Sandwich, Society, and Friendly Islands, who was tasked with keeping detailed records of commerce (MacAllan 2000:93). US presence in the islands strengthened relations with the monarchy and exacerbated the difficulties experienced by the British southern whale fishery, which in the short term led to abandonment of its bounty system and eventually to its withdrawal altogether. The dominance of the United States in Hawaii is easily visible in the numbers of shipping arrivals at Honolulu between 1824 and 1843 (Table 3.1). Of 1,680 whaling ships anchoring during that 20-year period, 1,377 were American—on average roughly 69 annually—while only 303 were British (Kuykendall 1931:49; Tate 1980:3). The prevalence of American whaleships of course meant that American merchant houses and shipping agencies established during the period of the China Trade were provided with great opportunities through the standardization of their visits.

Whaling agents in New England also recognized that standardization of seasonal visits and protection of American commerce in Hawaii could greatly benefit the industry. Realizing the potential to capitalize on the fruits of the preceding season's labor, the habit of transshipping cargos of oil soon became commonplace. As older whaling grounds were depleted, the need to venture ever farther into the Pacific, Indian, and Arctic Oceans grew. The cruises were lengthened even more by the inefficient habit of whaleships returning to their home ports in New England to unload their cargos and thus transiting back around Cape Horn into the Atlantic, which in turn resulted in

Table 3.1. Comparative Table for 20 Years of the Annual Arrivals of Whaling and Merchant Vessels at the Port of Honolulu, S.I., Formed from a Register Kept by Mr. S. Reynolds, Merchant of Honolulu

Whaling Vessels Visiting Honolulu	**1824**	**1825**	**1826**	**1827**	**1828**	**1829**	**1830**	**1831**	**1832**	**1833**	**1834**	**1835**	**1836**	**1837**	**1838**	**1839**	**1840**	**1841**	**1842**	**1843**	**Total**
American	69	22	91	64	84	83	75	60	101	89	93	63	57	50	62	58	41	53	53	109	1,377
British	18	14	16	18	28	26	19	21	17	18	18	13	16	17	11	2	4	6	14	7	303

Source: Wylie (1845:254).

time and money being lost. Thus, by the early 1830s, agents were establishing warehouses in foreign ports such as Honolulu as transshipment and resupply points (Davis and Gallman 1994:60–61). In an article about whaling in the North Pacific, Bockstoce (1984:531) explains that this practice cushioned financial hardship in the event of shipwreck and "also served to lessen the whaleship's function as a freighter to carry the ship's oil. If a ship were carrying the oil from an entire voyage, the captain might well have had to modify his tactics as the cargo accumulated, balancing the desire to preserve a valuable cargo with the need to take risks in search of whales. The practice of transshipping his cargo two or three time a year therefore allowed the master to maintain a more aggressive search for whales in waters that were often poorly charted or known to be dangerous."

The system of transshipping cargos proved to be one of the most important developments in nineteenth-century American whaling. Clipper ships that arrived in the islands with goods to be sold in merchant houses of Honolulu and Lahaina were soon returning to the mainland with oil and bone, while the whalers returned to their hunting (Hogue 1947:7). Voyages to the North Pacific and Arctic grounds soon averaged four years in length, though some lasted up to six. Whaleships themselves also participated in transshipment; if at the end of a cruise their holds were not full, the cargos were augmented with oil or bone from other vessels. Thus, this innovative practice resulted in a "much more efficient allocation of the industry's capital stock," and by the early 1840s, almost every agent adopted the process (Davis and Gallman 1994:60–61).

Throughout the first half of the nineteenth century, several ports in sheltered areas around the Hawaiian Islands developed to provide for visiting whaleships. Freshwater sources and stocks of firewood were abundant, and the climate proved perfect for agriculture and raising livestock. Communities prepared for the visits accordingly; farmers heavily planted the mountain slopes, ranchers fattened cattle and pigs, and wood was collected and stored. Aside from water and wood, visiting whalers could find freshly slaughtered meat; sugar, molasses, and coffee; vegetables such as potatoes, onions, taro, and cabbages; and fruits that helped to stave off scurvy, such as oranges, pineapples, melons, coconuts, pumpkins, onions, breadfruit, and bananas (Allen 1973:182). For whaleship crews who spent many months living on "salt-horse and hardtack," these fresh provisions were a welcome supplement to their diet (Day 1966:134).

A by-product of these seasonal visits was the introduction of Native Hawaiian sailors as a source of labor for whaleships. Often when the ships called at Hawaiian ports, they were missing crew members who had either deserted in other ports or had not survived the previous season. Whaling masters noted that Indigenous people of the islands were superb sailors and were "friendly, even-tempered, and generous"—all of which were characteristics that made them desirable additions to the crew of a whaleship (Allen 1975:183). The attractions of voyaging the world's oceans were enough to encourage many young Hawaiians to sign on to whaling cruises where they worked alongside international crews and experienced the world outside of their islands (Day 1966:134). Called *kanakas* from the Hawaiian word for "person," Hawaiians were expert crew members and soon became highly sought after for their skills, so much so that over time "New England whalers found it more lucrative to leave the U.S. with a skeleton crew and sign on 'kanaks' on their arrival in Honolulu" (Fischer 2002:101). Furthermore, their abilities were such that they quickly gained a reputation as being among the finest whalers and harpooners in the fleet, and by the 1830s as many as 3,000 Native Hawaiian seafarers were crewing American and European vessels (Fischer 2002:101).

While minor ports such as Waimea and Koloa on the island of Kauai or Hilo on Hawaii Island were convenient for obtaining fresh produce, recruiting local sailors, or simply for taking shelter (Gibbs 1977:60), it was Honolulu on the island of Oahu and Lahaina on Maui that became the major ports in the archipelago. Although port fees were much lower and prices of provisions reasonable, the anchorage at Lahaina offered poor protection from seasonal gales (Freidel 1943:381). As such, Honolulu dominated as the premier port throughout much of the whaling boom period (1820–1860). Offering a well-protected harbor, many shops, and ship repair facilities considered superior to any other in the North Pacific (Bradley 1942:217), Honolulu attracted the whaling fleets of the United States, UK, and many other nations (Gibbs 1977:60). In a relatively short time, the impacts of the whalers' retreats, however, took their toll; the waterfront lost much of its beauty and grew to look like "the seedier parts of ports back home, with boarding houses, gambling parlors, poolrooms, and grogshops in ample supply" (Dolin 2007:247).

By the middle of the nineteenth century, the ports of Honolulu and Lahaina were hosting hundreds of whaleships annually, peaking in 1846 when the number of arrivals reached 596 (Dolin 2007:246; Mrantz 1976:9). Undoubtedly, the phenomenal growth experienced throughout the Hawaiian Islands in the early to mid-nineteenth century was due in large part to the seasonal visits by whaling fleets. Hawaii's development as an entrepôt resulted from a combination of factors including the discovery of new hunting grounds, investment of capital by merchants, the standardization of seasonal stopovers, changes in the business structure of whaling ventures, and the way in which whaling agents capitalized on voyage profits. And while none of these can be seen as the single most important element, one thing was certain: it all depended on the success of the whaling crews while on the grounds and on their ability to access them safely.

Dangerous Path to Profit: The Northwestern Hawaiian Islands

The natural wealth of the Japan Grounds, in conjunction with others in the Central and South Pacific, was enough to supply the whale fleets with sperm oil cargos for nearly two decades. Eventually, however, those resources were overexploited, and the returns diminished, causing changes to the patterns of seasonal hunting. Instead of solely concentrating their efforts on the Japan Grounds, whalers passed through the area taking what sperm whales they could find before expanding their geographic scope to the north in search of the next boom. By that time, the numbers of British ships fishing in the North Pacific had all but disappeared, and the American fleet took a dominant role (Rydell 1952:65). Understanding that the waters to the north were generally too cold for sperm whales, they altered their strategy to also search for right whales.

These exploratory voyages quickly paid off; between 1830 and 1850, the whalers were rewarded with consecutive discoveries of several fertile right whale grounds across the Arctic (Kugler 1971:24), as well as bowhead grounds in the Bering Sea in 1848 (Bockstoce 2006:54). Coincidentally, during this period, prices for sperm oil dropped, and the demand for right whale oil and whalebone increased considerably (Bockstoce 1984:531). The new grounds soon became the focus of most of the American fleet's activities and remained so until the decline of the whaling industry around 1870. Throughout this period, Hawaii remained an important port of call for the whalers, but by the late 1850s, San Francisco slowly gained prominence before taking the place as the dominant Pacific port for Arctic whaling (Spears 1910:415).

Although much of the North Pacific was not explored officially or mapped until the middle of the nineteenth century, the lure of filling cargo holds with oil and turning quick profits brought whalers in droves. In doing so, the captains of these vessels greatly increased their chances of damage or loss through encounters with the many shallow reefs and tiny atolls that rise quickly from the region's generally deep water. Often, such dangerous oceanic features were completely uncharted, and when encountered at night or in conjunction with a storm, the results were catastrophic. Ships' captains did their best to fix the positions of each hazard when they stumbled upon them, but often, the reported positions were only relative and therefore of little help (Thrum 1915:134). Even when the positions were accurate, charts were seldom published during this period. Thus, the more prudent ships operating in the region maintained a constant and vigilant masthead watch to alert the helm of any danger sighted on the horizon (Friis 1967:262).

Though many of the potentially dangerous shoals and reefs that formed around Pacific islands and atolls were visible from a distance, others developed around remote seamounts, which made early detection nearly impossible. In many cases, isolated reefs sprang up out of deep, open-ocean environments; once charted, ships set courses to bypass them. In other cases, however, such features were not isolated and instead formed links in a chain that stretched over vast distances. These formations resulted in an increased chance of shipwreck, as vessels trying to avoid one reef often unwittingly encountered the next in the chain. The increase in maritime traffic in the Pacific in the early to mid-nineteenth century saw seemingly innocuous semisubmerged archipelagos transform into veritable ship traps. Throughout the nineteenth century, many merchant vessels and whaleships came to grief on islands, reefs, and shoals previously unknown to Americans or Europeans, which in turn received their Western names from the encounters (Van Tilburg and Kikiloi 2007:57).

Of the many remote formations that dot the Pacific, the area referred to as the Northwestern Hawaiian Islands (NWHI) are particularly relevant to nineteenth-century whaling. Located to the north of the main Hawaiian Islands, the NWHI run in a northwesterly arc for approximately 2,000 km and encompass a series of low islands, atolls, and barely submerged banks and reefs (McDowell Ward 2010:6–7). Most of the archipelago is characterized by deep water with intermittent areas of small, shallow reefs. Aside from the more substantial southern islands of Nihoa and Mokumanamana (Necker), which are known to have been occupied for hundreds of years but were abandoned by the time of Western contact (McDowell Ward 2010:11), few of the other islands within the chain could sustain human habitation for a lengthy period. While it is highly likely that voyagers and fishers utilized the region for marine resource extraction, by the time Europeans arrived, it is believed that the NWHI were not yet universally known to Native Hawaiians (PMNM 2011:19).

Western contact with the archipelago probably first occurred with the voyages of the French navigator Jean-François de Galaup de la Pérouse (Sharp 1960:151). A contemporary of Captain James Cook, La Pérouse's 1786–1789 expedition sighted and surveyed the island of Mokumanamana, which he called Necker Island in honor of the French minister of finance under Louis XVI (Gibbs 1977:175). He is also credited with identifying French Frigate Shoals, so-called for the type of ships under his command, as well as the 16.8 m tall rock named La Pérouse Pinnacle (Dodge 1971:36; Sharp 1960:151; Van Tilburg and Kikiloi 2007:57).

Soon after La Pérouse passed through the region, vessels engaged in the China Trade entered the southernmost part of the NWHI chain after provisioning and wintering at Kauai. In 1788, Captain James Colnett of *Prince of Wales* sighted the island of Nihoa, which he named Bird Island. The Spanish later captured Colnett's vessel off the West Coast of America; his detainment kept him from claiming the island and ultimately drove him insane (Rauzon 2001:8; Vancouver 1798:81–85). One of Colnett's contemporaries, Captain William Douglas of the English trading vessel *Iphegenia*, sighted Nihoa in 1789 and was for many years given credit for having first located it (Clapp et al. 1977:15; Meares 1791:28; Rauzon 2001:8–9; Thrum 1915:135). During the remaining years of the eighteenth century, a few other trading vessels passed near Nihoa en route to the Northwest Coast of America; however, none are known to have landed (Buck 1953:46; Clapp et al. 1977:15).

The next recorded European visit to the NWHI took place in 1805 with a Russian exploratory voyage under the command of Imperial Russian Navy officer Urey Lisianski. While surveying in the region, the Lisianski expedition ship *Neva* ran aground on a large, previously unrecorded shoal. After some effort, the ship was extracted from the reef and further exploration led to the identification of a nearby island. To mark its location, the island was given the name of the expedition leader, and the reef was named Neva Shoal after the ship (Clapp and Wirtz 1975:21; Gibbs 1977:175; Sharp 1960:189). From that time, few voyages by Westerners to the NWHI occurred until the advent of whaling on the Japan Grounds (Clapp et al. 1977:15).

While there is some speculation about the date of the first passage by a merchant vessel through the Japan Grounds (Richards 1999), undoubtedly it was the initial visit there by American and British whalers in 1820 that sparked the explosion in the numbers of ships entering the North Pacific and Arctic regions over the next 50 years. The stopovers in Hawaiian ports brought whaleships into contact with the NWHI. Whether sailing for the Japan Grounds, which stretched from Midway Atoll west to the Bonin Islands (Kugler 1971:24), or to those in the Arctic, the physical position of the NWHI became an obstacle to the success of the ships that dared enter.

In a short time, the uncharted physical hazards of the NWHI were exposed to the early whaling skippers who entered the region. Some of the vessels that encountered reef banks and shoals were fortunate to detect them from a distance and to lay down their locations on a chart. For instance, while on his second voyage to the Japan Grounds from the Sandwich Islands in 1820, Captain Joseph Allen of the whaleship *Maro* successfully identified and marked two such features: Maro Reef, a particularly treacherous location visible only by breakers (Robotti 1955:24; Van Tilburg 2002a:38), and Gardner Pinnacles, which he described as "a new island or rock not laid down on any chart[.] . . . 46 m high and about one mile in circumference[,] . . . it has two detached humps" (Clapp 1972:2; Stackpole 1953:269). Other vessels were not as lucky and instead spotted uncharted reefs too late, which resulted in their wrecking and their crews being stranded for sometimes lengthy periods. Such was the case for the British whaleships *Pearl* and *Hermes*, both of which were lost in 1822 on the extensive reef that now bears their name (Amerson et al. 1974:26; Sharp 1960:200). These examples illustrate two facts: first, that the NWHI contained many dangerous areas capable of destroying unsuspecting vessels; and second, that cautious navigation could result in safe passage through the region.

As crews gained knowledge of the particularly dangerous areas of the archipelago, it was disseminated among the whaling fraternity who fished the Japan Grounds. This knowledge transfer proved extremely useful; in the early to mid-nineteenth century, whaleships made thousands of successful passages though the region by charting courses that avoided the hazards. That is not to say that accidents did not still occur. Winter storms, poor judgment, and other circumstances led to occasional losses in the NWHI throughout the nineteenth century, which in turn contributed to their lasting reputation as being almost impossibly dangerous. An example of this public sentiment is found in a contemporary popular publication that listed the maritime casualties in the archipelago. In *The Hawaiian Annual and Almanac: The Reference Book of Hawaii*, Reverend J. M. Lydgate (1915:133) suggests that "the islands and reefs to the northwest of Hawaii have been a veritable graveyard of marine disaster. The two sufficient reasons for this have been, first, the low, inconsistent character of the islands, and second, the faulty or insufficient location of them on the marine charts. The menace of an iceberg is the fact that it lies seven-eighths under water, and you strike some submerged, protruding spur of it before you dream of danger. In a much more disastrous way the same thing is true of many of these islands."

Regardless of their reputation, the potential for quick returns and profit far outweighed the risks of passing thorough the NWHI, and by the late 1820s, over 100 whaleships were successfully fishing on the Japan Grounds each year. For the most part, the risks were significantly lessened through industrial strategies, such as establishment of fishing patterns that seasonally shifted between the North and South Pacific. Using knowledge of whale migration as a guide, ships were able to make the most of the North Pacific whaling season and then safely withdraw before the arrival of the dangerous winter weather. In some cases, however, whaleships remained on the grounds too late in the season or were simply caught off guard by inclement weather or miscalculations in navigation—11 American or British whaleships are reported to have directly encountered the reefs of the NWHI between 1822 and 1867. Of those 11, only 1 was fortunate enough to have been refloated and saved, while the other 10 became total losses. Chapter 5 includes more detailed information pertaining to the construction, crews, wrecking events, and postwrecking activities of the American ships.

Decline of Importance for American Whaling and Hawaii

The golden age of whaling ushered in a new era for the Hawaiian archipelago, which bolstered its growing economy through the purchase of provisions, goods, and services (Dolin 2007:246–247). By 1870, however, the significance of the islands to the Pacific whaling fleets had diminished. As had occurred with the Japan Grounds, identification of new whaling grounds caused a major shift in both industrial tactics and structure.

Just as whalers shifted their focus from the Pacific sperm whale grounds to those of the right and bowhead whales in the Arctic, so did the New England whaling firms relocate their warehouses and transshipment agencies. Thanks in large part to the California gold rush, San Francisco grew as an important American maritime center and, in a relatively short amount of time, became the primary base of operations for Pacific whalers (Ellis 1991:163–167). Other factors affecting Hawaii's decline as a major port for whaleships essentially mirrored those that led to the decline of whaling in general. These factors included overexploitation of whale populations, the shift from whale oil to petroleum and other types of oil, increased outfitting and operating costs, and the decimation of the whaling fleet during the American Civil War (Kugler 1971:26; Mrantz 1976:35).

Nevertheless, unlike the whaling industry, Hawaii's economy was able to survive. As the numbers of ships visiting the islands each year declined, entrepreneurs in Hawaii turned their focus toward expanding sugarcane and rice production. The climate and soil on the islands proved perfect for large-scale farming of these and other agricultural commodities and was given a boon through a treaty of reciprocity with the United States in 1875 (Freeman 1951:346–347). The capital generated from increased agricultural production was reinvested in the industry, and plantations were quickly established throughout the islands. The success of the sugar industry also enabled Honolulu to once and for all emerge as Hawaii's most important shipping port, while its competitors from the whaling days languished (Ellis 1991:163).

The wrecks that occurred in the NWHI also had some direct effects on the development of the Hawaiian Islands, though for the most part these can been viewed as positive. The simple fact that ships were lost on reefs to the northwest had bearing on transfer of knowledge about the region. As hazards were encountered, they were ascribed names in English and (when possible) their positions were recorded. As soon as possible, that information was disseminated among the whaling community to prevent any further losses. Evidence of the effectiveness of this word-of-mouth knowledge transfer can be seen in the fact that out of the hundreds of ships that passed through the archipelago between 1822 and 1842 only three were lost on its dangerous reefs.

While the relative locations of some of these reefs and shoals occasionally found their way onto nautical maps, for the most part they were manually added on charts kept by individual captains. Eventually, however, the wrecks of American whalers that occurred in the NWHI—and throughout the Pacific—led to the undertaking of official mapping expeditions. Though many shippers and whalers lobbied the government to send a US naval vessel into the region to conduct an examination of "the coasts, islands, harbors, reefs and shoals of the Pacific Ocean and South Seas" (Friis 1967:263), it was not until the newspaper editor Congressman J. N. Reynolds became involved that the government finally acknowledged the need to create accurate nautical charts of the region (Boggs 1938:184; Friis 1967:263; Leff 1940:5–7). Between 1839 and 1842, an American expedition led by US Navy (USN) Captain Charles Wilkes carried out extensive scientific research and mapping operations in the Pacific Ocean, with the main objective of aiding the whaling industry (Dodge 1965:54; Freeman 1951:87). The Wilkes expedition produced excellent charts that were further refined through other voyages into the NWHI over the next few decades; these included the 1859 USN expedition by John M. Brooke on USS *Fenimore Cooper* and the exploratory commercial voyage of Captain N. C. Brooks in the Hawaiian bark *Gambia* (Brooke 1986; Brooks 1860; Day 1966:171).

Conclusion

Undoubtedly, the economic impact of whaling on the Hawaiian Islands was responsible for its transformation into a bustling entrepôt by the mid-nineteenth century. From the earliest visits by whalers, the archipelago's convenient location in the Central Pacific, the subtropical environment, abundance of fresh produce, and excellent deepwater anchorages provided a perfect sanctuary for reprovisioning and rest after months or years at sea. The industrial shift in operational focus to the Pacific Ocean saw small ports grow quickly into centers of trade and activity. This development, in turn, resulted in changing business strategies that significantly impacted the ways in which the American fishery operated and forever changed the Hawaiian Islands.

4

Graveyard of Marine Disaster

The History and Archaeology of American Whaleships Lost in the Northwestern Hawaiian Islands

Although many whaling ships are known to have wrecked throughout the Pacific between the 1820s and 1890s, their remote loss locations keep most beyond the reach of archaeologists. Notable exceptions are those deposited in the NWHI, where of the 10 whaleships that came to grief on its shoals and reefs during that period, 5 have been identified. This chapter offers an overview of the region's physical environment and status as a marine protected area, as well as a summary of the activities of managers and researchers to locate and document its maritime heritage. Basic descriptions of eight of the whaleships wrecked in the region are provided, along with an overview of the circumstances that led to their sinking, the rescues of their crews, and (where applicable) archaeological surveys. Finally, it describes the extensive historical and archaeological investigations of the wrecks of the American whaleships *Two Brothers* and *Parker*, which are the focus of discussion in following chapters.

Papahanaumokuakea Marine National Monument

Several wrecked whaleships are located within the boundaries of Papahanaumokuakea Marine National Monument (PMNM or the monument). The PMNM encompasses the NWHI chain, the major emergent features of which include Nihoa and Necker Islands, French Frigate Shoals, Gardner Pinnacles, Maro Reef, Laysan Island, Lisianski Island, Pearl and Hermes Atoll, Midway Atoll, and Kure Atoll (Rauzon 2001:2). Together these islands, atolls, and reefs have the distinction of being the world's most remote archipelago, as well as one of its most significant marine protected areas (PMNM 2008:6).

Situated in a vast, remote, and largely uninhabited region of the central Pacific Ocean, the NWHI were set aside as a marine protected area called the Northwestern Hawaiian Islands Marine National Monument in 2006 under Presidential Proclamation 8031. The original boundaries originally extended roughly 1,931 km in length by 185 km in width and cover 362,062 square km of the Pacific Ocean (PMNM 2008:2–5). In 2007, the title was changed to Papahanaumokuakea Marine National Monument to signify the importance of the area to Native Hawaiian culture, with the name reflecting "the ancient Hawaiian tradition that relates to the birth or formation of the Hawaiian Islands as is personified by the earth" (Van Tilburg and Kikiloi 2007:56). The 2016 Presidential Proclamation 9478 expanded the boundaries of PMNM, which quadrupled its size to 1,508,870 square km and created the largest protected area on Earth (Chan 2020:1) (Figure 4.1).

The oceanic features of the NWHI region include "shallow coral reefs, deepwater slopes, banks, and seamounts" and support an array of marine life (PMNM 2011:1). Due to their remote oceanic location, the atolls of the NWHI are considered one of the most pristine marine ecosystems in the world (Pandolfi et al. 2005; Smith 2010:3) and provide refuge and habitats for a wide range of threatened and endangered species (PMNM 2011:1). Though the reefs were impacted by roughly 150 years of human exploitation, for the last half century, the ecosystems have been in a recovery mode and "intact populations now characterize the predator-dominated reefs" (Friedlander and DeMartini 2002; Kittinger et al. 2011:2). To date, over 7,000 species of fish, birds, marine mammals, invertebrates, and algae have been described in PMNM, though some researchers believe that the region's true biodiversity is far greater (Fautin et al. 2010; Toonen et al. 2011:2).

Ecosystem protection of this area has a long history, which began when US President Theodore Roosevelt established what is now known as the Hawaiian Islands National Wildlife Refuge to protect its natural resources in 1909 (Kittinger et al. 2011:2; PMNM 2011:1). Since that time, conservation efforts in the region have strengthened through the establishment of other federal protections including the Northwestern Hawaiian Islands Coral Reef Ecosystem Reserve, Midway Atoll National Wildlife Refuge (NWR), the Battle of Midway National Memorial, and the Hawaiian Islands NWR (McDowell Ward 2010:6). The state of Hawaii has also acknowledged the significance of the region through the designation of the Northwestern Hawaiian Islands State Marine Refuge (Department of Land and Natural Resources [DLNR] 2005; Kittinger et al. 2010:209). The most recent recognition of both the natural and cultural uniqueness of this place came in 2010 with the United Nations Educational, Scientific and Cultural Organization (UNESCO) World Heritage Committee's unanimous inscription of PMNM as a mixed site—the

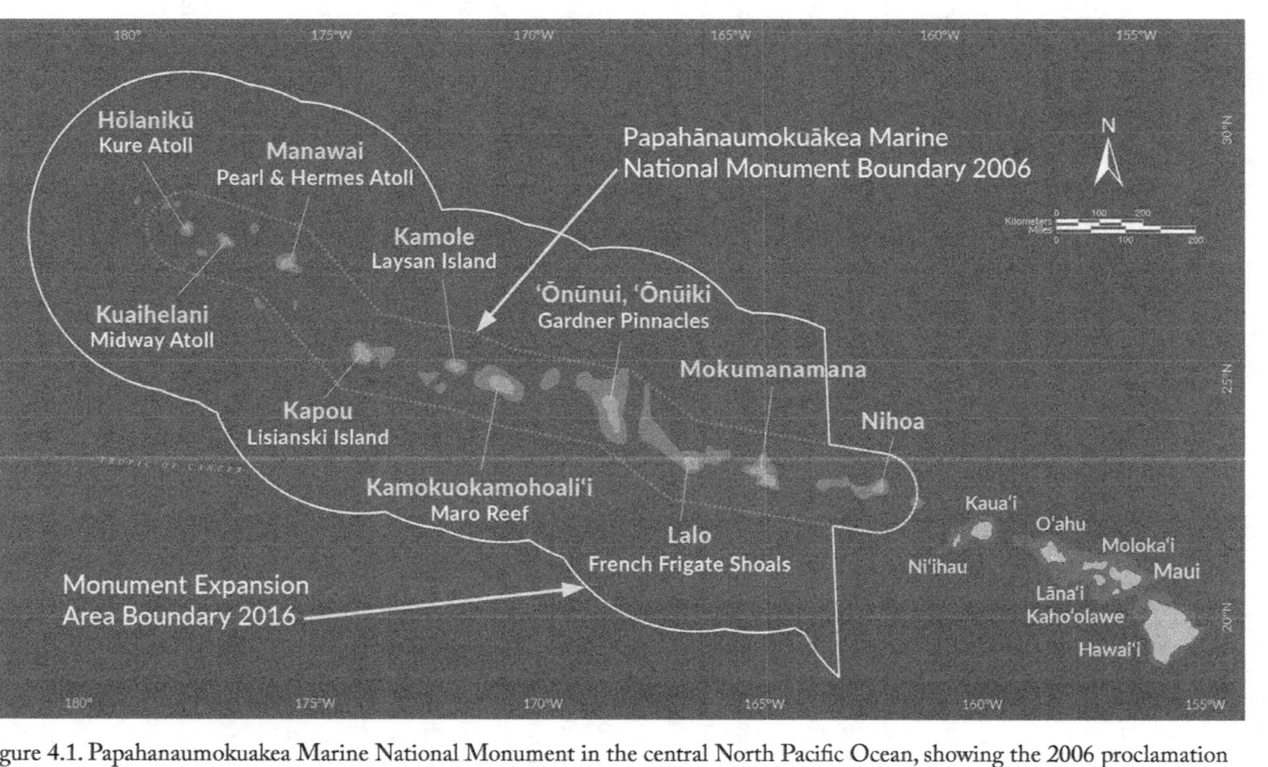

Figure 4.1. Papahanaumokuakea Marine National Monument in the central North Pacific Ocean, showing the 2006 proclamation boundary and the 2016 expansion boundary. (Courtesy of Papahanaumokuakea Marine National Monument.)

first of its kind in the United States and one of only 27 such sites in the world (PMNM 2011:1).

Management of PMNM is the responsibility of three cotrustees: the state of Hawaii Department of Land and Natural Resources (DLNR); the US Fish and Wildlife Service (FWS); and the National Oceanic and Atmospheric Administration (NOAA). These agencies work cooperatively to fulfill the tasks of "preserving the ecological integrity of the Monument and perpetuation of the NWHI ecosystems, Native Hawaiian culture, and historic resources" (PMNM 2008:1). Protocols for management of this marine protected area were outlined in a memorandum of agreement between the cotrustees, whose policies and regulations are enforced in conjunction with applicable state and federal laws (Kittinger 2010:209).

The heritage of PMNM represents hundreds of years of continuous utilization of the NWHI. Polynesian voyages of exploration, commercial ventures during the ages of sail and steam, fortification and battles during World War II, fishing activities, and even present-day research expeditions have all left reminders to help tell the story of human interaction with the archipelago. The monument's diverse heritage includes oral histories, chants, and archaeological sites associated with Native Hawaiian voyaging and use of the region, as well as many terrestrial and submerged sites related to the colonial and historic periods (PMNM 2011:1). To best understand and preserve these diverse and complex cultural remains, they are divided into three categories for management: Native Hawaiian culture and history, which includes both tangible and intangible heritage relating to the deep genealogical, cultural, and spiritual connections of Native Hawaiians to the region (Van Tilburg and Kikiloi 2007:58); historic resources, which include historic period remains located on the various islands; and maritime heritage resources, which include submerged and beached shipwrecks, aircraft, and many other sites of historic, cultural, and archaeological significance (PMNM 2011:1–5; US Fish and Wildlife Service 2008).

Whaling Research in PMNM

At least 10 American and British whaling ships came to grief on the barely submerged reefs and shoals within the boundaries of PMNM between 1822 and 1867 (Table 4.1). Primary source data relating to these ships and their losses led to the rediscovery of five of those wrecks. Their archaeological investigations offer an opportunity to learn more about vessels that operated in the Pacific during the whaling industry's golden age (Raupp and Gleason 2010:66).

The first archaeological surveys in the NWHI were conducted in 1998 by the University of Hawaii Marine Option Program (UH MOP). Those were

Table 4.1. Reported Wrecks of Whaling Vessels Occurring in the Northwestern Hawaiian Islands, 1822–1867

Vessel Name	Nationality	Home Port	Tonnage	Date Built	Date Lost	Approximate Location	Captain
Pearl	British	London	320	1805	1822	Pearl and Hermes Atoll	Clark
Hermes	British	London	262	1811	1822	Pearl and Hermes Atoll	Taylor
Two Brothers	American	Nantucket	217	?	1823	French Frigate Shoals	Pollard
Gledstanes	British	London	428	1817	1837	Kure Atoll	Brown
Parker	American	New Bedford	406	1831	1842	Kure Atoll	Sherman
Holder Borden	American	Fall River	442	1841	1842	Lisianski Reef	Pell
Konohassett	American	Sag Harbor	426	?	1846	Lisianski Reef	Worth
South Seaman	American	Fairhaven	497	1856	1859	French Frigate Shoals	Norton
Daniel Wood	American	New Bedford	345	1852	1867	French Frigate Shoals	Tallman
Unidentified pre-1859	American	Unknown	?	?	Pre-8/30/1859[a]	Laysan Island	?

[a] Wreck seen and reported by Captain N. C. Brooks of the Hawaiian bark *Gambia* on August 30, 1859.

followed by NOAA-supported UH MOP research in 2002, as well as investigations by maritime archaeologists employed by NOAA's Office of National Marine Sanctuaries in 2003. These projects resulted in identification of several shipwreck sites, including that of the American whaleship *Parker* (1842) in 2002. Since that time, surveys have been conducted by researchers from PMNM, UH MOP, the Hawaii Institute for Marine Biology, Flinders University, and East Carolina University, as well as other government agencies and archaeological consulting firms.

Due to high expedition costs and time constraints imposed by conditions in the open ocean environment of the region, starting in 2003 researchers consulted with marine scientists from other disciplines and marine debris cleanup crews to understand indicators of possible shipwrecks and to gather information pertaining to previously sighted cultural materials. This approach proved successful and resulted in identification of several sites, including two wrecked whaleships, the British whalers *Pearl* (1822) and *Hermes* (1822). Archival records and historic maps indicating the locations of wrecks helped to determine prospective survey areas and eventually led to identification of two other whaleship wrecks, the British ship *Gledstanes* (1837) and the American ship *Two Brothers* (1823). Overall, when considering the immensity of the area contained within PMNM boundaries, the identification of half of the known vessels of a single type is quite an accomplishment.

Pearl (1822) and *Hermes* (1822)

The first-reported occurrence of lost whale hunters in the NWHI claimed two vessels, *Pearl* and *Hermes*, and resulted in the reef on which the wrecks occurred being endowed with their names. On April 4, 1822, the two British whaleships departed Honolulu together and set a course for the newly noted Japan Grounds (Hawaiian Gazette 1868). Sailing on the night of April 24, and unaware of the dangers that might be encountered in the unexplored region, both ships ran aground on a previously uncharted reef approximately 1,600 km to the northwest of Honolulu (Spoehr 1988:80). Accounts state that Pearl was the first to hit the reef and became hard aground. Seeing its mast light and noting distress, *Hermes* attempted to look after its consort but unfortunately met a similar fate (Galtsoff 1931:49; Van Tilburg 2002a:21). The two wrecked approximately 16 km from one other and were unable to be extricated; *Pearl* apparently became wedged after running head-on into the reef, while *Hermes* was likely pushed broadside onto it (Morrell 1832:217).

No lives were lost from either of the ships, and after the storm abated, the survivors set up camp and salvaged as much of the provisions and materials from the wrecks as possible. Realizing that the likelihood of rescue by another ship was slim, they developed a plan to construct a small schooner from the wreckage of the two ships (Amerson et al. 1974:26; Van Tilburg 2002a:21). Just as the castaways prepared to launch the schooner, which they named *Drift*, the British whaler *Thames* came to their rescue and removed all but twelve of the crew, who elected to stay (Galtsoff 1931:50). Together, the remaining whalers sailed the newly built schooner on a successful 10-week voyage back to Honolulu (Amerson et al. 1974:26; Spoehr 1988:80). After recuperating in the islands, most of the crews of *Pearl* and *Hermes* made their way back to England working onboard other whaling or merchant ships. Two of the *Hermes* crew, however, decided to remain in Hawaii, where they established a highly successful ship repair business that catered to whaleships (Taylor 1952).

Several historic accounts and an 1859 chart by Captain N. C. Brooks describe the wrecks of the whaleships *Pearl* and *Hermes* and provided researchers approximate locations for each of the shipwreck sites (Brooks 1860). In 2002, a team from UH MOP identified the remains of the whaleship *Pearl* on the southeastern edge of Pearl and Hermes Atoll. Subsequent investigations over the next five years yielded extensive cultural materials and a complete map of the site (Van Tilburg 2002a, 2005, 2006, 2007). From these studies, researchers determined that the ship indeed encountered the outer reef head-on and became wedged in an indentation of its spur-and-groove topography before breaking apart (Thomas 2006:3). Although wooden structural elements such as the keel and hull planking were noted as buried in the sandy seabed, the site is largely characterized by metal, brick, glass, and ceramic artifacts situated both outside the reef and just inside the lagoon. Among the artifact assemblage are anchors, cannon, try-pots, tryworks bricks, tryworks knees, pintle, hawsepipe, hull sheathing, a grindstone, fasteners, and fragments of bottles and dishes (Raupp and Gleason 2010:70–71; Van Tilburg 2005). All artifacts documented at the site support identification of the wreck as that of an early nineteenth-century British whaleship.

While conducting tow boarding surveys along the outside of the southern portion of the atoll, NOAA marine debris crews identified artifacts consistent with a shipwreck site (Van Tilburg 2004). Further investigation by archaeologists from the NOAA Maritime Heritage Program confirmed the remains as those of the whaleship *Hermes*. Situated in an extremely dynamic area, the site is characterized by concentrations of material culture nestled in pockets over a large area of the reef. Artifacts documented at the site included anchors, cannon, cannonballs, musket shot, bricks, ballast, a bronze spigot, a blubber hook, gudgeon and pintle, hawsepipe, fasteners, and bottle fragments (Van Tilburg 2005). Begun in 2005, site recording was completed in 2008 and resulted in a detailed site map (Raupp and Gleason 2010:71). As with *Pearl*, all documented artifacts are consistent with those expected from an early nineteenth-century British whaleship.

Gledstanes (1837)

While cruising in what he thought to be open ocean, the captain of the London-based ship *Gledstanes* sighted a distant island that was "not layed down on any chart" (Hawaiian Spectator 1838:336). Though a strong current was running, he allowed the vessel to drift under reduced sail, thinking it would bypass the island. At around midnight on June 9, 1837, however, the watch called out that the reef was under the vessel, and soon it was surrounded by breakers (Sandwich Island Gazette 1837). *Gledstanes* struck the northern side of a reef that was later determined to be part of Kure Atoll, and within a few days in the heavy surf, it was a total loss (Casserley 1998:60; Sandwich Island Gazette 1837).

Though one intoxicated crew member died after jumping overboard into the surf, the rest of the crew were able to launch three whaleboats and eventually made landfall on the previously sighted island on the far side of the atoll (Hawaiian Spectator 1838:336). After establishing a makeshift camp, the crew picked up pieces of wreckage with the intention of constructing a vessel. Using shipbuilding tools crafted from salvaged whaling spades and lances on a makeshift forge, the crew built a small vessel, which they named *Deliverance*, within three months of their wrecking (Sandwich Island Gazette 1837; Woodward 1972:4). The captain and a small crew then sailed to the Sandwich Islands while the rest of the party remained on the island for several more months before being rescued in a vessel sent by the British Consul in Honolulu (Casserley 1998:60; Read 1912:7; Woodward 1972:4).

As with the search for other early wrecks in the NWHI, Captain N. C. Brooks's 1859 accounts of surveys of the region's geographic features and accompanying maps proved instrumental to the identification of the wreck of *Gledstanes* (Brooks 1860). Using the Brooks chart of Kure Atoll as a guide, divers drifting along the unusually calm outer reef quickly spotted the vessel's remains in 2008. The site is scattered over a discrete area of the reef and consists mainly of large iron artifacts settled into channels of the spur-and-groove topography. Although some objects were obscured due to scouring, most were easily identifiable; these include anchors, cannon, anchor chain, iron ballast bars, the lower portion of a try-pot, tryworks bricks, and copper fasteners. Analysis of this assemblage suggests use onboard a mid-nineteenth-century sailing ship involved in whaling activities, most likely the British whaleship *Gledstanes* (Raupp and Gleason 2010:71).

Holder Borden (1844)

On November 6, 1842, the relatively new vessel *Holder Borden* departed Fall River, Massachusetts, for its first whaling voyage (Daily Mercury 1845). After 17 months of successful hunting in the South Atlantic and South Pacific, *Holder Borden* stopped at Honolulu to refit and recruit. Seven days after leaving, the ship cruised to the northwest when the winds began to strengthen and forced the crew to reduce sails (Friend 1844). By the early morning hours of April 12, 1844, gales developed, which drove the ship onto a sandbank, and shortly thereafter, it swung around and hit a coral reef from which it could not be extricated (Clapp and Wirtz 1975:22; Dupont 1954:357). At daylight, the crew found that they were wrecked upon an enormous shoal, approximately 6.5 km from a small sand island (Chandler and Phillips 1848:194). After attempting unsuccessfully to free the vessel, the crew salvaged provisions, cut away the masts (to keep the vessel from falling over), launched the boats, and rowed for the island (DuPont 1954:358).

Though they wrecked on Neva Shoal and camped on Lisianski Island, neither was on the ship's charts. Deeming it to be previously undiscovered, it was called Pell's Island after the *Holder Borden*'s captain (Friend 1844). Fresh water and rations were quickly salvaged from the wreck and seals, turtles, and seabirds sustained the crew while they were stranded (Casserley 1998:16; Clapp and Wirtz 1975:22). With provisions secured, the captain kept the crew busy by salvaging everything possible, including anchors, cables, sails, provisions, and clothing, as well as 1,400 of the 1,800 barrels of oil that were onboard (Friend 1844). A plan was soon hatched to build a small schooner from the wreckage. To do so they constructed tools from the ship's fittings in a forge built using tryworks bricks and fired by coal onboard. They crafted saws from barrel hoop iron and made a box for steaming planks out of a try-pot (Dupont 1954:362–363). In a few months, the completed vessel was painted, sheathed, copper-fastened, and named *Hope* (Polynesian 1844). Soon thereafter, Captain Pell and 24 of the crew sailed it to Maui, where the schooner was sold. After purchasing a larger vessel, the brig *Delaware*, the captain returned to rescue the remaining crew and the salvaged goods that had survived (Polynesian 1844; Whalemen's Shipping List 1845).

As noted above, the shoal on which *Holder Borden* wrecked was not known to its captain. It had been identified, however, in 1805 when the Russian exploring vessel *Neva* ran aground on the adjacent reef. After narrowly escaping the near wrecking event, the ship's name was ascribed to the reef, and the island was named for its captain, Urey Lisianski (Clapp and Wirtz 1975:21). Based on the description of the vessel's loss provided in historic accounts, as well as a map drawn by Lieutenant John M. Brooke (1859) indicating the wreck, a probable loss location at Neva Shoal was established. Due to the mazelike structure of the extensive coral reefs, attempts to identify the wreck of *Holder Borden* have so far proved unsuccessful.

Konohassett (1846)

Only two years after the loss of *Holder Borden*, the American ship *Konohassett* became the next whaler to wreck in the NWHI. As with *Holder Borden*, the former merchant vessel *Konohassett* left its home port of Sag Harbor, New York, on December 6, 1845, bound for the Pacific and its first whaling voyage. Taking no whales on the outbound leg of the trip, the ship stopped at Lahaina to recruit before sailing to the northwest for unspecified grounds (Friend 1846). Apparently also unaware of the existence of Neva Shoal, *Konohassett* was cruising under full sail in the early morning of May 24, 1846, when it struck a reef approximately 27 km southeast of where *Holder Borden* wrecked only two years earlier (Adams 1918:337; Clapp and Wirtz 1975:22). The impact of increasing swells quickly resulted in the ship being bilged, and the crew was forced to leave the wreck. The next morning, they returned to the ship and from that vantage point sighted the island; realizing the ship was a total loss, they proceeded to shore where they found the remains of the *Holder Borden* survivor camp (Clapp and Wirtz 1975:22).

Understanding that their only chance of survival was through self-rescue, Captain Worth determined that they needed to construct a vessel. For four days, the crew salvaged provisions and any useful materials and tools from the wreck. On May 28, construction began and after only 18 days a "fast-sailing sloop," which they named *Konohassett Jr.*, was complete (Friend 1846). Though they experienced some difficulty keeping it watertight, the sloop carried the captain and six crew members to the port of Honolulu where they arrived on July 31, 1846 (Finckenor 1975:44; Friend 1846). The American Consul in Honolulu dispatched the Hawaiian schooner *Halileo* on August 4, 1846, to Pell's Island to rescue the remaining crew (Clapp and Wirtz 1975:23; Ward 1960:63–67).

The historic description of the loss location "17 miles south-east of where *Holder Borden* wrecked only two years earlier" provided an approximate area for archaeological surveys (Adams 1918:337; Clapp and Wirtz 1975:22). This position corresponds with one marked for the wreck on Captain N. C. Brooks's 1859 chart of Lisianski Island (Brooks 1860). Situated among the maze of reefs at the extreme southern edge of Neva Shoal, the survey area requires extended diving surveys to ensure that all patch reefs are searched. To date, the wreck of the whaleship *Konohassett* has not been identified.

South Seaman (1859)

Thirteen years passed before the next whaleship became a casualty of the NWHI. The clipper-style ship *South Seaman* of Fairhaven, Massachusetts, departed Honolulu on March 6, 1859, for a whaling cruise to northern grounds and then to the Sea of Okhotsk (Atlas and Daily Bee 1859; Daily National 1859; Saturday Press 1883). Though the location of French Frigate Shoals was well-known and charted by this time, an error in navigation led the ship's captain to "suppose himself to be full forty miles to the westward" (Atlas and Daily Bee 1859). While sailing in a stiff wind in the early morning hours of March 13, 1859, the masthead lookout spotted breakers, and almost immediately, the ship struck a coral reef at French Frigate Shoals with such force as to break the main royal mast on impact (Pratt 1859). The ship then pounded heavily on the bottom until it was forced on top of the reef; there it rested on its side, the bottom completely smashed and the breakers making a clean sweep through it (Atlas and Daily Bee 1859). At daylight, the order was given to abandon ship, but before taking to the whaleboats, the crew saved nautical instruments, charts, provisions, and materials to use for shelter (Pratt 1859; Walker 1909:16).

All hands onboard were saved, but when the boats cleared the breakers, there was no land in sight; the decision was made to sail for the Ladrone Islands (Guam), which, although farther away than Honolulu, were in the track of the trade winds (Boston Daily Journal 1859; Lydgate 1915:136; Pratt 1859). After only two hours at sea, the schooner *Kamehameha IV*—fortunately sealing and guano hunting in the area—spotted the whaleboats and rescued them. Due to limited space onboard, some of the crew were left on a nearby island, while the captain and others were transported to Honolulu (Amerson 1971:39; Atlas and Daily Bee 1859; Pratt 1859). The wreck and stores were sold at auction to the owners of *Kamehameha IV*, which soon returned to collect the remaining survivors and salvage *South Seaman* (Pacific Commercial Advertiser 1859, 1867a; Van Tilburg 2002a:42).

Historic records relating to the loss of *South Seaman* provide an approximate location of its place of wrecking at French Frigate Shoals. Of particular interest is a map drawn in 1859 by Captain N. C. Brooks that illustrates a break in the reef that is dubbed South Seaman Passage and includes a sketch thought to represent that vessel (Brooks 1860). Using such historic sources as a guide, researchers consulted modern charts and determined an estimated location for the shipwreck. Attempts to survey this area of the fringing reef over several field seasons failed to relocate the site, often owing to large swells and unsafe operating conditions.

Daniel Wood (1867)

The American barque *Daniel Wood* was the last historically known whaleship wrecked in the NWHI. After hunting on an unspecified ground to the west, the New Bedford–owned vessel arrived at Honolulu on April 5, 1867; discharged a small amount of oil for transshipment; and departed five days later to continue whaling "to the northward" (Pacific Commercial Advertiser 1867a). Sailing in clear weather on April 14, 1867, the barque was running a course to pass between Necker Island and French Frigate Shoals. Though the captain had earlier determined their position to be clear of any danger, throughout the day strong and unpredictable currents altered the ship's course (Lydgate 1915:136; Schwemmer 2004). Just after midnight, the masthead lookout sighted breakers, and, despite efforts to avoid it, *Daniel Wood* struck French Frigate Shoals and drifted into the reefs where it heeled over in powerful surf (Daily Mercury 1867). With no chance of saving the vessel, the masts were cut away and the boats launched; the crew rowed until they reached the nearest land. The following day they returned to the barque and salvaged all possible supplies from the wreck, which they stored on shore.

By the next day the vessel had broken up, and no remnants of it were visible above the water (Lydgate 1915:136; Ward 1960:553–563).

Captain Richmond immediately determined that the best chance of rescue lay with an expedition back to Honolulu. Though challenged by a lack of tools, the crew selected their best whaleboat, salvaged wreckage, and proceeded to build up its sides and deck it over, creating what sailors refer to as a "sister gunwale and washboard" (Lydgate 1915:136; Pacific Commercial Advertiser 1867a). In only two days, the vessel, called *Anne E. Wilson*, was completed and fitted out with some provisions. Leaving 27 crew members with the rest of the supplies on the island, the captain and seven crew embarked on the 450-mile (724 km) journey to Honolulu (Lydgate 1915:136). After resting briefly and resupplying at the island of Niihau, the crew arrived in Honolulu. There they met the USS *Lackawanna*, which was immediately dispatched to rescue the remaining survivors (Pacific Commercial Advertiser 1867a; Ward 1960:553–563).

Historical accounts for the loss of *Daniel Wood* provide little indication for its point of contact on French Frigate Shoals. Information provided in a contemporary newspaper article about the wreck suggest that the crew rowed approximately 24 km from the wreck until they reached a "small sand bank, barren, with the exception of here and there a tuft of grass" (Pacific Commercial Advertiser 1867a; Ward 1960:553–563). A later article from the same newspaper reported that it was sold at auction and that the buyers recovered valuable materials from the wreck "in deep water outside the reef" (Pacific Commercial Advertiser 1867b). Using this information, a survey of the proposed survey area was undertaken in 2012. Unfortunately, heavy swells resulted in abandonment of offshore surveys, and those inside the reef failed to identify any remains (Green and Raupp 2012:6–7).

Unidentified (Pre-1859)

Only one other whaling ship was reported lost in the NWHI, though little historical information about it is available. The only known reference comes from the summary account of Captain N. C. Brooks's 1859 survey in the Hawaiian bark *Gambia*, which was looking for potential guano deposits in the archipelago. In that report, Brooks includes a list of 11 shipwrecks and the islands on which they came to grief (Brooks 1860; Casserley 1998:4). Of those 11, one is initially recorded as being a "ship, name unknown, on Laysan Island," but supplementary data indicates that "the wreck at Daysan [Laysan], the name of which I was unable to ascertain, was that of an American whaleship" (Ward 1960:309–310). While the chart that Brooks (1860) created for Laysan Island includes the relative location of a "unknown" shipwreck, it is uncertain whether it represents this vessel, and no archaeological surveys have been undertaken.

Two Ships on the Rocks

Of the six American whaleships known to have wrecked within the boundaries of PMNM, only two have been located and thoroughly documented. These sites represent some of the earliest remains of American pelagic whaleships found and archaeologically investigated anywhere in the world. Thanks to protections provided by both their remote geographic positions and their inclusion within the boundaries of a marine protected area, these relatively untouched sites perhaps offer an opportunity to elicit information about the technologies and industrial processes employed, as well as the social and cultural conditions that existed, onboard American whaleships of this period.

Two Brothers

The first American whaleship known to have wrecked in the NWHI was the Nantucket vessel *Two Brothers*. Lost at French Frigate Shoals in 1823, just less than a year after the wrecks of the British ships *Pearl* and *Hermes*, it, too, was a victim of shallow reefs encountered while attempting to traverse the uncharted region. The *Two Brothers* shipwreck site has the historic distinction of being directly connected to one of the most storied episodes in the industry's history—the loss of the whaleship *Essex*. This tragic incident involved the sinking of a ship rammed by a whale in a remote part of the Pacific Ocean and later became the inspiration for Herman Melville's epic *Moby Dick* (1850). The same captain that operated *Essex*, George Pollard Jr., was later the captain of *Two Brothers*. Archaeologically, the site is distinct as it represents the earliest known wreck of an American whaleship identified in the Pacific region and provides the most complete artifact assemblage of any whaleship lost in PMNM. Also of interest is the existence of multiple survivor accounts of the wrecking event, which aided in the positive identification of the wreck. Together, these factors make *Two Brothers* one of the most important and extensively documented shipwrecks located so far in the monument.

History of the Whaleship *Two Brothers*

Unlike many of the other ships that came to grief on the reefs of the NWHI, little is currently known about the construction or working life of *Two Brothers* prior to its loss at French Frigate Shoals. The few features of the ship that are known were gleaned from an entry in Alexander Starbuck's *History of the American Whale Fishery* (1878). The Nantucket-owned *Two Brothers* first appears in that register in 1818, where it is described as 217 tons, ship-rigged, and captained by George B. Worth (Starbuck 1878:226). With no owner or agent listed, the ship sailed for the Pacific Ocean on September 25, 1818, and returned on October 20, 1820, with a cargo of 378 barrels of "Sperm-oil" and 1,836 barrels of "Whale-oil" (Starbuck 1878:227). Though no data about the actual cruise are recorded, the much greater cargo of whale oil than sperm oil might suggest that they encountered large numbers of right whales while passing through the South Atlantic, possibly on the Brazil Banks, and hunted there for at least part of the cruise. Though the object of the cruise may have been the more lucrative sperm whale oil, during this period it was common to take right whales when they were seen since "cargoes of the cheaper (whale) oil could be obtained much more quickly, with shorter voyages and equal profits" (Richards 1994:23). The same 217-ton ship *Two Brothers* appears only one other time in Starbuck's table; in 1821 it is listed as having been "lost on a coral reef" while under the command of George Pollard Jr. (Starbuck 1878:236–237).

Using Starbuck's data (1878:227–237) regarding the size and rig of the ship as a guide, efforts were made to identify the origin, build date, and any other construction data pertaining to *Two Brothers*. While the name appeared to be popular for vessels of all classes during this period, only one matched the known dimensions listed by Starbuck: the 217-ton ship *Two Brothers* built at Hallowell, Maine, by Joseph Glidden in 1804 (Baker 1973:929; Barker 1879:9). Regional shipbuilding records indicate that this vessel measured 25.73 m in length, 7.44 m in beam (width), and 3.72 m in depth. They also state that its original home port was Gardiner, Maine, that its first owner was S. Bradstreet, that the first master was James Purrington, and that it was fitted with a female figurehead (Baker 1973:659–929). The dimensions of this vessel fit well with those desired for whaleships of this period, and many vessels employed by Nantucket whaling operations were constructed in Maine shipyards, which "turned out many superior ships for the trade" (Martin 1975:45). Thus, it is possible that this record belongs to the same *Two Brothers* that was lost in PMNM in 1823.

If this Maine-built ship was in fact the same *Two Brothers* commanded by Captain Pollard, as yet no information about the vessel is known pertaining to its career between the years 1804 and 1817. While it is possible that it was employed in the whale fishery during that period, it seems unlikely since no mention of it is found in Starbuck's comprehensive shipping returns table for the period (Starbuck 1878:202–225). Instead, it is probable that, like many other New England ships, it was laid up during the period of Jefferson's embargo and the War of 1812 (Baker 1973:186) and then later used in some branch of the merchant service before being sold to Nantucket owners to be refit as a whaler.

The next known historical mention of *Two Brothers* comes from its association with the whaleship *Essex*. This now-famous incident involved the harrowing experience of the crew of an ill-fated whaleship, twice rammed by an angered whale while hunting in an open stretch of the Pacific Ocean, roughly

2,000 km northeast of the Marquesas Islands (Whipple 1979:85). According to an account of the loss (Chase 1821), the whale's second sortie resulted in the hull being stove at the bow and the ship quickly taking on water (Dakin 1934:88). Salvaging what little provisions they could access, Captain George Pollard Jr. and the 20 crew members set out in three whaleboats hoping to make it to the coast of South America. Though efforts were made to keep the three small boats together, storms soon separated them. After surviving a traumatic 94-day journey that resulted in sickness, starvation, and ultimately cannibalism, the remaining crew in two of the boats were rescued at sea within five days of one another, while the third boat was never heard from again (Whipple 1979:87).

The five survivors of the *Essex* tragedy were taken to Valparaiso, Chile. There they recuperated before four of the crew returned home aboard the Nantucket whaleship *Eagle*. Deemed too weak to accompany the crew, Captain Pollard remained a further two months in Chile before gaining passage to Nantucket aboard the whaleship *Two Brothers* (Philbrick 2001:193). Pollard's integrity so impressed the ship's captain over the course of the two-and-a-half-month return voyage that he was recommended as a replacement master of the whaleship that brought him home (Philbrick 2001:203). The freakish nature of the accident that led to the loss of *Essex* did little to tarnish Pollard's reputation as a fine skipper, and he soon found himself preparing for yet another voyage to the Pacific, this time in command of *Two Brothers* (Heffernan 1981:145).

In late November 1821, little more than three months after his return to Nantucket, Pollard sailed for the Pacific Ocean as master of *Two Brothers* (Macy 1835:249). Interestingly, Pollard was not the only *Essex* survivor onboard; as a sign of their trust in his skills, former crew members Thomas Nickerson and Charles Ramsdell chose to again serve under him (Philbrick 2001:206). After a long and stormy yet uneventful passage through the Atlantic, the ship rounded Cape Horn and eventually arrived safely in Talcahuano, Chile, where it met the Nantucket whaler *Martha*. There, the captains of the two vessels agreed to "throw their chances together and cruise for whales far to the westward" (Nickerson n.d.). This hunting strategy appears to have been commonly used by American and British whalers in the early years of the Pacific fishery and involved two ships agreeing to "mate," or unite, forces in a temporary partnership (Brown 1887:260; Fonda 1969:11; Kugler 1980:7; Vickers 1985:282). After each took on recruits in that port, the two ships met off the coast of Peru and cruised west-northwest on a course for the recently identified Japan Grounds, possibly even passing the spot where the whale struck *Essex* only two years earlier (Heffernan 1981:149).

By early February 1823, *Two Brothers* and *Martha* turned more toward the north and were cruising in roughly the same latitude as French Frigate Shoals when the weather deteriorated (Nickerson n.d.). Despite only being able to rely on dead reckoning due to overcast skies, both captains judged their position to be well west of any danger and maintained their courses (Philbrick 2001:208). The two ships were barely visible to one another when, on the afternoon of February 11, heavy squalls and severe gales caused the crew of *Two Brothers* to reduce sail. Soon they saw large breakers, and though they attempted to change course, "the high sea running behind" them affected the ship's steering and almost immediately it struck on a reef (Nickerson n.d.). Though the vessel briefly refloated, it quickly struck the reef again with such force that it shattered the stern (Gardner 1823). In his account of the wrecking event, boatsteerer Thomas Nickerson stated that there were "breakers apparently mountains high" and that soon the stricken ship was being pushed over onto its broadsides and pounded so hard "that one could scarcely stand upon his feet" (Nickerson n.d.). The ship was beaten to pieces, and water filled it so quickly that the pumps were of no use (Macy 1835:249). Nickerson (n.d.) noted that while "Captain Pollard seemed to stand amazed at the scene before him," two of the mates stepped in and ordered the masts be cut away in an attempt to save the ship. Pollard soon came to and realized that *Two Brothers* was beyond saving; knowing that the falling masts and spars would likely damage the whaleboats, he ordered the crew to drop their axes and instead to prepare to abandon ship (Nickerson n.d.; Philbrick 2001:209). Although two whaleboats were each loaded with four oars, a sail, and some navigational equipment, no fresh water, provisions, or spare clothing could be obtained (Gardner 1823). The entire complement of officers and crew escaped in the crowded boats as the ship was smashed in the heavy surf (Nickerson n.d.; Philbrick 2001:209).

The crew of *Two Brothers* passed the night tossed around in tempestuous seas, and eventually, the boats became separated. In the morning, one of the boats noticed a sail in the distance; they rowed toward it and found it to be their companion, *Martha*. Though this ship, too, had run aground on a reef after parting an anchor cable during the storm, quick action by the chief mate freed it with little damage sustained (Nickerson n.d.). The other boat of *Two Brothers* crew saw a high rock in the distance and pulled for it through the night but, upon reaching it, found the area to be inhospitable. Instead, they rowed to the south where the following day they encountered three small islands, one of which they were able to land on and take refuge. They established a makeshift camp and began to search for food, but soon, one of the men saw a sail on the horizon, which proved to be *Martha* (Gardner 1823). They rowed the whaleboat out to meet the ship and with the crews of both whaleships onboard, *Martha* sailed for the island of Oahu. They arrived three weeks later, and after a brief rest, the ship soon departed for Nantucket

(Macy 1835:250). There is no indication of attempts to salvage the wreck of *Two Brothers*.

Site Location and Research Expeditions

During tow boarding operations at French Frigate Shoals in August 2008, maritime archaeologists noted a previously unidentified ship's anchor on the edge of a reef in the extreme northwest corner of the shoals. Extending the search out from the anchor, researchers found the remains of a wooden ship scattered in pockets of the surrounding reef and spilling over its sides. Preliminary analysis of the artifacts identified suggested that they belonged to an early nineteenth-century whaleship. As discussed earlier in this chapter, only three whaleships are known to have wrecked at French Frigate Shoals: *Two Brothers* (1823), *South Seaman* (1859), and *Daniel Wood* (1867). Though the physical evidence strongly indicated that the remains belonged to the earlier of the three, no definitive proof of the vessel's identity was found, so the wreck was given the working name Shark Island Whaler, in reference to a nearby sandy island. The site and all artifacts were then surveyed thoroughly, mapped spatially, and recorded photographically as part of the initial site inspection.

The next visit to the Shark Island Whaler site occurred in 2009 and resulted in the unanticipated detection of a second portion of the wreck. This chance find was a by-product of ecological research conducted by a graduate student from the University of Hawaii, the purpose of which was to determine whether "shipwreck sites have distinct ecosystems and if there are residual effects from shipwreck disturbances that manifest as differences between wreck and surrounding coral reef communities" (Smith 2010:4). To collect data needed to complete a thesis project, previously identified shipwreck sites were investigated in detail using several predetermined methods. Once work at the wreck site was completed, the same methods were applied to a randomly selected control area of "comparable size and habitat composition approximately 100 m away from the wreck plot on the contiguous reef" (Smith 2010:5). In the case of the Shark Island Whaler site, the selection of the control location proved serendipitous, and divers quickly noted shipwreck artifacts. Closer inspection revealed remains of the same vintage to those recorded on the Shark Island Whaler site (hereafter referred to as Section A). Thus, the area was deemed to be a second part of the same site (Section B), and researchers quickly recorded as much data as possible from the new section in the time available.

Maritime archaeologists and volunteers from PMNM and Flinders University returned to Section B of the Shark Island Whaler site in 2010. A large area of the reef was surveyed visually, and the scattered remains were recorded thoroughly using multiple methods including spatial mapping, artifact recording, and photography. The information gleaned from this investigation led to the determination that the remains were those of the Nantucket whaleship *Two Brothers*. As such, specific efforts were made to identify and record as many diagnostic artifacts as possible in the hope of making a conclusive identification. A permit was also obtained that allowed for the recovery of a small number of diagnostic artifacts from the site for conservation, analysis, and later public display at PMNM's Mokupapapa Discovery Center in Hilo, Hawaii. Marine remote sensing was also undertaken in the area surrounding the reef and island to identify any associated anomalies or debris. A magnetic contour map and side-scan sonar mosaic for the surveyed areas were created, and numerous anomalies were detected (SEARCH 2010:71–79).

After thorough analysis and consideration of all evidence recovered from the two sites, the identity of the site as that of the whaleship *Two Brothers* was formally announced on February 11, 2011, the one hundred eighty-eighth anniversary of its loss. In the meantime, PMNM and the Nantucket Historical Association agreed to create a museum exhibit highlighting the importance of the wreck to both the NWHI and the island of Nantucket. After obtaining a permit for artifact recovery, a PMNM maritime heritage team visited Section B of the *Two Brothers* site in August 2010 to retrieve three harpoon tips, two lance tips, two ceramic sherds, and a cast-iron cooking pot. Upon completion of conservation of these objects, the exhibit, titled *Lost on a Reef*, was installed at the Nantucket Whaling Museum and opened in the summer of 2012.

Site Description

The shipwreck of the whaleship *Two Brothers* is situated in the extreme northwestern portion of French Frigate Shoals near a small sandy islet named Shark Island. Positioned among the shallow reef complex immediately to the east of the island, the site measures approximately 100 m in length by 150 m in width. Depths on the site range from 1 to 8 m depending on tides, and visibility generally ranges from 10 to 20 m. The reef on which the wreck is located is typical of those in French Frigate Shoals and includes abundant marine life and coral species, including rare large table corals known as *Acropora* (Maragos and Gulko 2002:32). Coral encrustation on the cultural material associated with shipwreck sites is generally moderate to heavy.

The shipwreck comprises two parts separated by a relatively wide channel between two reefs. The first of these, Section A, stretches over an area of reef that measures approximately 120 m in length by 100 m in width (Figure 4.2). It is characterized by cultural materials associated with an early nineteenth-century wooden sailing vessel engaged in pelagic whaling, which are scattered mainly in pockets in the reef that extend onto the reef flat to the north. Diagnostic evidence supporting the vessel's function includes anchors, rigging components, an abundance of copper rather than iron fasteners, and hawsepipes, as well as the bases of two trapezoidal-shaped bottles. Artifacts specifically associated with whaling include three try-pots and hundreds of tryworks bricks. No wooden structural elements of the ship were noted, and they are unlikely to be present on the site; due to the dynamic subtropical environment of the region, it is probable that such remains would have either floated away as the ship broke apart and/or were eaten by wood-boring organisms over time.

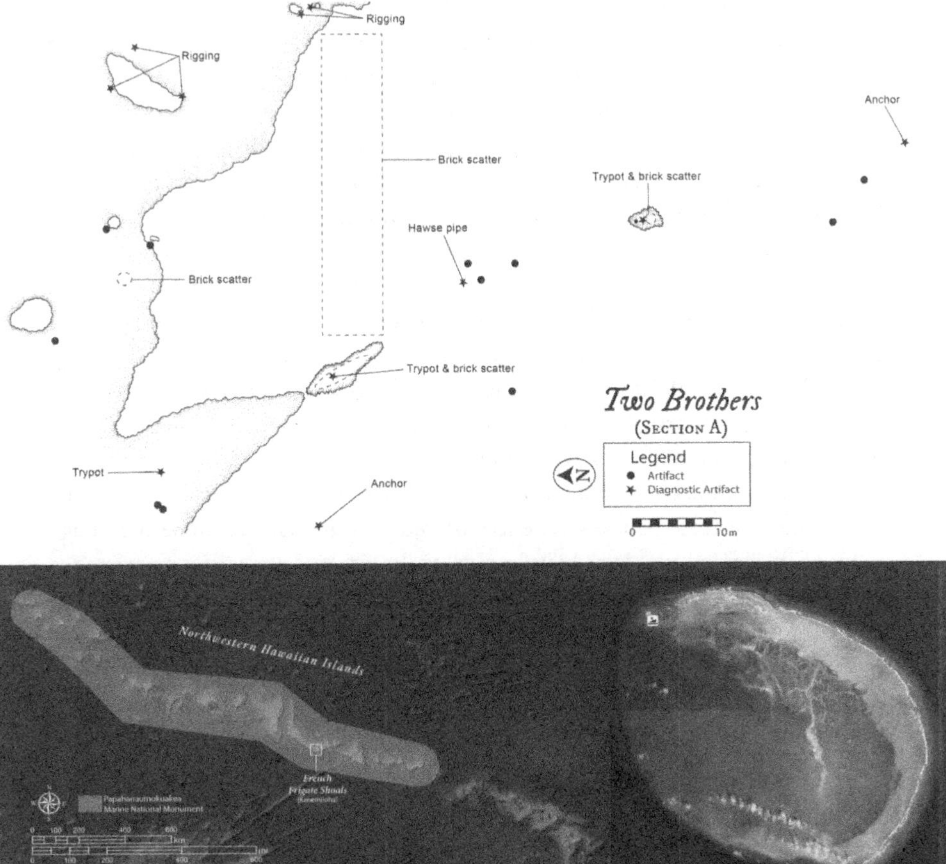

Figure 4.2. Site plan for Section A of the *Two Brothers* shipwreck site at French Frigate Shoals. (Courtesy of Jeremy Borrelli.)

Section B of the *Two Brothers* site is located approximately 200 m to the south of Section A. This section is much smaller than the other and measures approximately 50 m in length and 30 m in width. While it can also be characterized by whaleship artifacts located on top of and in pockets of a central reef structure, much of the diagnostic cultural materials are found on the reef flat to the south or in a large, sloping depression toward the northern part of the reef. Many of the diagnostic artifacts identified indicate an association with the aft section of a whaleship, including a possible kedge anchor, a grinding stone, ceramics, galley wares, several items of whalecraft (tools used in hunting whales), and possible components of navigational equipment. As with Section A, no wooden elements of the ship were noted during investigations of the site.

Evidence for Identification as *Two Brothers*

While archaeological data indicated that the Shark Island Whaler site indeed contained the wreck of *Two Brothers*, two main issues warranted consideration before a positive identification of the site could be made. The first is the possibility that the remains represent another, completely undocumented whaleship that wrecked on that reef. While this possibility is always a concern when dealing with shipwrecks found in remote locations, the fact that the crews of contemporary ships lost in the same region all survived those wrecking events and that their losses were duly reported by several news outlets make an unreported shipwreck an unlikely conclusion. Instead, it seemed far more probable that the site is in fact the remains of *Two Brothers* and that the historically reported location of its loss was incorrect. To better understand this issue, the available accounts of the loss of *Two Brothers* were scrutinized along with other historical information pertaining to it.

By all accounts, the captains of both *Two Brothers* and *Martha* were convinced that the reef their vessels encountered was uncharted. Since the geography of the region was largely unknown at the time of wrecking, its existence was not questioned; instead, its position was dubbed Two Brothers Reef and added on nautical charts. Some doubt about the veracity of their positioning, however, is found in a letter penned by boatsteerer and shipwreck survivor Thomas Nickerson to author Leon Lewis dated October 1876. Nickerson described discussions he had with Thomas Derrick, the first mate of *Martha*, in which they both agreed that the reef on which *Two Brothers* was lost was in fact French Frigate Shoals, "notwithstanding our two Captains believed and reported that this was a new discovery" (Nickerson 1876). The two mates alleged that navigation errors due to nearly two weeks of poor weather prevented the captains from taking accurate lunar observations to determine their position. Instead, they were forced to rely solely on dead reckoning (Nickerson 1876; Philbrick 2001:208).

The assumption made by Nickerson and Derrick was eventually verified through the voyages of several vessels that surveyed the region over the next century. While it is probable that other whaleships passing the area attempted to validate the supposed location of Two Brothers Reef, the first known record of such activity occurred through an 1859 USN survey of the region by Lieutenant John M. Brooke in USS *Fenimore Cooper*. In his journal of the voyage, Brooke referred to it as "Brothers" reef, and, though he expressed doubts about its existence, he nevertheless passed over the assigned position "without perceiving any indication of land or shoal" (Brooke 1986:43). Later that same year during pioneering investigations of the NWHI, Captain N. C. Brooks of Honolulu crossed over the given position in the bark *Gambia* but, finding no shallow reef, determined that *Two Brothers* struck on French Frigate Shoal (Brooks 1860:500; Hydrographic Office 1903:145). Subsequent attempts to identify the reef also proved fruitless; these include efforts by USS *Albatross* in 1902 (US Coast and Geodetic Survey 1919:51) and USS *Tanager* in 1923 (King 1931:16). As only "great depths were obtained at its reported position" (US Coast and Geodetic Survey 1919:51), the location for Two Brothers Reef was eventually stricken from official charts. The fact that none of these attempts to verify the position given by Captains Pollard and Pease for the loss of *Two Brothers* was successful, when combined with modern hydrographic and satellite imaging data that indicate no shoals in the region, provides strong support for the theory that the whaleship was indeed lost on French Frigate Shoals.

Clues to the ship's identity were also found using the accounts of first mate Eben Gardner and boatsteerer Nickerson, two of the shipwreck's survivors. Details extracted from those accounts provide a provocative image of the wrecking event and the destruction of the ship and offer spatial data that was used to analyze the site layout and distribution of artifacts at the shipwreck site. For instance, Gardner stated that after initial impact on the reef, the hull briefly refloated before it "struck again so heavy and shattered her whole stern" (Gardner 1823). To this description, Nickerson added that the ship was soon "careening over upon its broadsides and thumping heavily" (Nickerson n.d.). When taken together, these two descriptions indicate the ship broke apart and scattered over a large area of the reef, spilling the contents of the ship as it moved.

Upon completion of archaeological recording of the site, plan-view illustrations of each section were made. When georectified, these two plans matched the information put forth in the survivor accounts. The approach from the south, leading to the wreck, is deep enough to allow a ship the size of *Two Brothers* to easily pass until it encountered the inner line of reefs, which could be the location of the ship's initial impact. Because the depth of this inner reef is not shallow, it is probable that the "mountainous seas" described by

Nickerson (n.d.) could have refloated the ship after grounding. If the survivor's accounts are correct, the following sea that affected the ship's steering (Nickerson n.d.) would have also pushed them directly into the next reef, which is much shallower and roughly the distance of "once the length of the ship," as posited by Gardner (1823). Furthermore, as described above, Section B of the *Two Brothers* site (Figure 4.3) encompasses a relatively small area and is characterized by artifacts associated with known use areas of a whaleship's aft section, such as the galley and bosun's locker (Stackpole 1967:31). Thus, Section B likely represents the stern of *Two Brothers* that is described as shattering in the second impact with the reef.

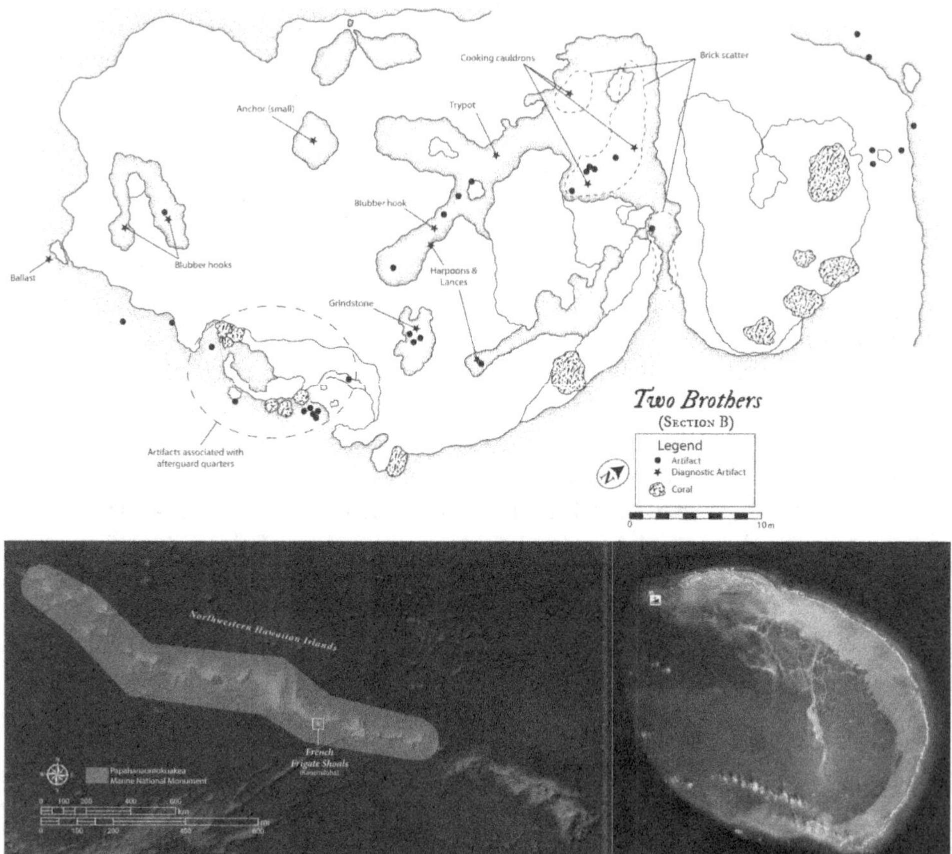

Figure 4.3. Site plan for Section B of the *Two Brothers* shipwreck site at French Frigate Shoals. (Courtesy of Jeremy Borrelli.)

The physical remains documented at Section A of the Shark Island Whaler site also match Nickerson's assertion that the ship was soon "careening over upon its broadsides and thumping heavily" (Nickerson n.d.). The composition of the reef complex in that area is such that the sections of reef holding the remains are separated by a relatively deep, sandy channel. As the artifacts found scattered across the reef to the north indicate that they were likely associated with the midships and forward parts of the vessel, it is possible that Nickerson described the motion of the ship after losing its steering from the smashing of its stern. Uncontrollable, the vessel could have bounced along until reaching the channel, where it began to list before being pushed onto the subsequent reef where it spilled open and foundered. This supposition is further supported by the fact that numerous rigging components were recorded on the reef flat on the northern side of the outer reef, likely indicating that the masts of the vessel came to rest there.

The historical information pertaining to the wrecking of the whaleship *Two Brothers* and the physical descriptions provided by survivors present a solid case for the identity of the shipwreck. Furthermore, the archaeological deposits found at the site are indicative of those associated with an early nineteenth-century wooden ship employed in pelagic whaling. While it is true that no particular object can be viewed as a "smoking gun," as with other shipwreck identifications that have recently come under scrutiny, "the preponderance of the evidence supported the wreck's identity . . . and often, with archaeology, that is as good as it gets" (Wilde-Ramsing and Ewen 2012:113).

Parker

After the loss of *Two Brothers*, nearly 20 years elapsed before another American whaleship wrecked in the NWHI. In autumn of 1842, the New Bedford whaleship *Parker* became a total loss after encountering the fringing reef at Kure Atoll during a squall. The approximate location of the loss, wrecking event, and subsequent stranding of the surviving crew on Ocean Island (now Green Island) were detailed in an 1843 account from the shipkeeper. In 2002 staff and students from the University of Hawaii relocated the remains of *Parker*. The site is characterized by numerous artifacts scattered over a large area of the lagoon (Van Tilburg 2002a). It is archaeologically significant as it provides evidence of the advances in ship technologies employed by American whaling interests as the fishery developed in the first half of the nineteenth century. As with the *Two Brothers* shipwreck, the account of the wrecking and subsequent activities of the survivors were integral in the positive identification of the wreck.

History of the Whaleship *Parker*

The whaleship *Parker* was a typical vessel of the American whaling fleet in the 1830s. Information pertaining to its construction was compiled by the Survey of Federal Archives, an initiative by the Works Progress Administration (WPA) tasked with collecting data on all vessels registered through the federal Customs Districts (WPA 1940). According to the program's publication titled *Ship Registers of New Bedford, Massachusetts* (1940), it was built at Fairhaven, Massachusetts, in 1831 by master carpenters Fish and Delano and jointly owned by Fredrick Parker, John A. Parker, Hayden Coggeshall, and Joseph Dunbar. Purpose-built for the whaling trade, the 406-ton ship is described as having two decks, three masts, a square stern, no galleries, and a billethead on the bow. The ship measured 34.2 m in length by 8.7 m in beam and had a depth of 4.3 m when it was surveyed by L. M. Allen and registered at New Bedford on October 6, 1831 (WPA 1940:248). Interestingly, measurements for the ship found scribbled in two places in the back of the log of *Parker*'s inaugural voyage differ slightly from those of the original survey; while the length and breadth dimensions match closely, the statement "depth of hold from sealing [*sic*] to under space of upper deck beams 19 ft~6 in [5.9 m]" indicates a difference in depth of approximately 1 m (Brown 1835:229). This discrepancy likely resulted from the point at which the measurements were taken. Though the hold of most ships was generally considered to be the intended cargo space in the lowest part of the ship, whaling captains used all available interior space for oil storage, and thus, the definition of *Parker*'s hold was likely extended to include the between decks area.

Prior to its loss at Kure Atoll in 1842, the whaleship *Parker* successfully completed two lucrative cruises under the ownership of J. A. Parker & Sons (Starbuck 1878:282–283). The first of these voyages departed on the day of its initial registration with the Customs District at New Bedford. In his "Table Showing Returns of Whaling Vessels from American Ports," Alexander Starbuck (1878:282–283) lists the ship as bound for the Pacific Ocean under the command of Charles F. Brown on October 6, 1831. Over the next three years, *Parker* hunted on unspecified grounds and called at Honolulu at least four times, transshipping approximately 3,800 barrels of oil back to New Bedford (Van Tilburg 2003:32). On February 24, 1835, the whaleship returned to New Bedford with a profitable cargo of 3,150 barrels of sperm oil (Starbuck 1878:282–283).

The log for this voyage survived and is now part of the Nicholson Whaling Log Collection at the Providence Public Library in Providence, Rhode Island. Aside from the daily observations and comments expected of a ship's log, additional data includes details of provision management onboard, exact specifications for parts of the ship and whaling equipment, a tally of whales taken per boat, and an extensive table that recorded air and water temperature, wind direction, and weather observations made during its passage from the equator to the Pacific Ocean (Brown 1835:248–252). Included among the entries on the final pages of the log are coordinates for a right whale ground east of New Zealand, which were provided by the captain of the whaleship *Golconda*, the same for a sperm whale ground west of the Marquesas Islands given by the captain of the ship *Lancaster*, and the general comment "Dusky Bay New Zealand is said to be a good place for right whales in June and July" (Brown 1835:254).

With the success of *Parker*'s inaugural voyage, its owners wasted little time in outfitting the ship for a return to the Pacific Ocean. Under the command of Master William Austin, *Parker* departed New Bedford on May 30, 1835, and remained at sea for four years. Though the log for this voyage is unknown, New Bedford whaling agent Dennis Wood provided a summary of the ship's movements in his unpublished manuscript "Abstracts of Whaling Voyages of the United States 1831–1873" (Wood [1873]). From Wood's description, the aforementioned information noted in the back of the log of the previous cruise influenced the decision of where to hunt. Over the course of the voyage, *Parker* spent some time on sperm whales around Tahiti but most often reported hunting the waters off New Zealand and calling at its ports (Wood n.d.:397). This strategy proved productive, and on May 3, 1839, the ship sailed into its home port laden with a mixed cargo of 1,523 barrels of sperm oil, 1,539 barrels of whale oil, and 15,200 pounds (6,894 kg) whalebone (Starbuck 1878:316–317; Wood n.d.:397).

The continued success of *Parker*, and the seemingly endless number of whales found in the Pacific, led owners J. A. Parker & Sons to organize another trip soon after its return. Under the command of Captain Prince Sherman, the whaleship departed New Bedford on August 26, 1839, for a 40-month voyage with orders to "cruise for 25 months for sperm, but if unable to procure 2,000 barrels in that time to proceed to the N.W. for right whale" (Temperance Advocate and Seamen's Friend 1843). According to the entry in "Abstracts of Whaling Voyages of the United States 1831–1873," much of the early part of the voyage was spent hunting sperm whales off the South American coast (Wood n.d.:397). In December 1841 *Parker* was cruising the On-the-Line grounds around the equator when its captain drowned after an encounter with a large sperm whale. At that point, first officer George M. Smith succeeded him in command of the ship and decided to call at Lahaina to recruit before continuing the voyage (Temperance Advocate and Seamen's Friend 1843).

Upon leaving Hawaii in April 1842, Captain Smith followed the orders of *Parker*'s owners and steered a course for the northwest coast. The ship spent the following months hunting right whales—likely on the newly discovered (1835) Northwest Grounds off the coast of what is now British Columbia (Van Tilburg 2003:32)—and then departed for Hawaii around the first of August (Wood n.d.:397). It appears that their course led them onto the eastern part of the Japan Grounds since the Wood abstract for this cruise lists an encounter with the ship *Geo. Champlin* on September 15, 1842. Only eight days after this sea meeting, the weather deteriorated, and *Parker* was "lost on a ledge of rocks" (Wood n.d.:397).

The shipkeeper, Richard F. Quinn, provided an extensive account of the wrecking event, stranding on Ocean Island, and later rescue of the survivors. First reported in the Hawaiian newspaper Temperance Advocate and Seamen's Friend on June 27, 1843, various sources in New England used direct transcriptions of that account to report on the wreck (Ward 1960:505–533). According to Quinn, on September 23, 1843, *Parker* encountered squalls and rain from the north to northeast that continued to build throughout the day. Despite having shortened nearly every sail to weather the storm, the ship drifted until Ocean Island was sighted to the southeast. At 2:30 on the morning of September 24, "a sea dashed through the cabin windows of the *Parker* and she struck the reef 8 miles N.N.W. from the centre of the island" (Temperance Advocate and Seamen's Friend 1843). Within an hour of impact, it was clear that the ship was a total wreck, so members of the crew cut away the masts before abandoning it. Four members of the crew were lost in the ensuing struggle, but the other 22 were able to cling to a makeshift raft of floating masts and spars and eventually cross the reef. After eight days and seven nights of suffering and fighting intense currents to remain in the lagoon, the survivors reached Ocean Island (Temperance Advocate and Seamen's Friend 1843).

Though the survivors were physically debilitated after the ordeal, they understood the severity of their predicament and the reality that it would likely be some time before they could be rescued. The crew searched the beaches and found some wreckage of the British ship *Gledstanes*, lost at Kure Atoll in 1837, which served as firewood and construction material for a simple shelter (Woodward 1972:4). Although only a few provisions were salvaged after *Parker* broke up, the island provided plenty of food; over the next six months, the castaways estimated that they consumed over 7,000 seabirds, 60 monk seals, and a dog left by the crew of *Gledstanes* (Temperance Advocate and Seamen's Friend 1843). Food was prepared using utensils fashioned from pieces of copper salvaged from *Parker*'s wreckage (Van Tilburg 2003:33; Woodward 1972:4). The newspaper account states that a constant lookout was maintained and that the captain held small religious services daily (Temperance Advocate and Seamen's Friend 1843).

On April 16, 1843, signals were made to a sail seen on the horizon, which proved to be the New Brunswick ship *James Stewart*. On the following day, Captain Smith and three other survivors were taken aboard and offered passage to Honolulu. A quantity of provisions and some other useful articles were provided to the remaining 20 members of *Parker*'s crew, and the captain of *James Stewart* pledged to return for them as soon as possible. On May 2, 1843, however, the remaining survivors were rescued by the New Bedford whaleship *Nassau* and transported to Honolulu. Thirteen survivors remained there under the protection of the US Consulate, while the other seven signed on with Captain Weeks to serve as crew on *Nassau* (Temperance Advocate and Seamen's Friend 1843).

At the time of its wrecking, the whaleship *Parker* had onboard 2,000 barrels of sperm oil and 1,000 barrels of whale oil, all of which was lost (Starbuck 1878:354–355). The ship was insured for a total of $55,800 under separate policies issued by five different companies based in New Bedford and New York. When combined, the value of these policies and the market value of the cargo onboard amounted to $82,000 (Whalemen's Shipping List 1843). There is no record of any salvage of the wreck, and the only remains known to have been recovered are those utilized by the shipwreck survivors.

Site Location and Research Expeditions

Maritime archaeologists first examined the wreck of the whaleship *Parker* in 2002. Working as part of a larger multidisciplinary research expedition known as the Northwestern Hawaiian Islands Coral Reef Assessment and Monitoring Program (NOWRAMP) (Van Tilburg 2002a), the team from the University of Hawaii documented the site while conducting baseline cultural heritage surveys of Kure Atoll (PMNM 2011:11; Van Tilburg 2002a:1). While investigating a reported anchor resting in shallow water in the northwest section of the lagoon, they instead found the remains of what appeared to be a mid-nineteenth-century shipwreck. Recorded artifacts include three anchors, an anchor chain, copper fasteners, iron strapping, mast hoops, deck machinery parts, copper sheathing, a brick, pipes, pieces of lead, and a variety of encrusted unidentified artifacts (Van Tilburg 2002a:54). Though limited by time constraints, an initial measured sketch of the site was completed and the artifact scatter documented photographically (Van Tilburg 2003:33). Analysis of the artifacts indicate that the site likely represented the bow section of a ship pushed over the reef crest (Van Tilburg 2002a:54). Although no positive identification of the site was made at the time, artifact analysis suggested that the remains were most likely those of either the British whaleship *Gledstanes* or the American whaleship *Parker* (Van Tilburg 2002a:54).

In 2003, a team consisting of maritime archaeologists returned to the site to conduct further investigations (Van Tilburg 2003:1). Their primary focus was delineating the site boundaries by visually surveying and recording areas inside the lagoon that surround the artifact concentration recorded in 2002, as well as conducting marine magnetometer surveys outside the reef. Visual surveys resulted in detection of an extensive scatter of fasteners, machinery, anchors, chain, rigging elements, hull sheathing and bricks, and a ship's bell, all stretching in a line from the inside of the reef crest to the northwest. Magnetometer surveys detected numerous potential magnetic anomalies in the area outside the reef corresponding to the scatter trail; limited testing of these identified only a small number of fasteners. The results of the survey expanded the boundaries of the site and reinforced the theory that the remains represent those of a mid-nineteenth-century whaleship that collided with the reef and was subsequently pushed over into the lagoon, dropping artifacts in a trail as it broke apart. While no conclusive identification for the wreck was made, data analysis indicated *Parker* as being the most likely candidate (Van Tilburg 2003:33).

In 2005 and 2006, teams from NOAA returned to accurately record the spatial distribution of the scattered site (Van Tilburg 2006). To achieve this, major artifacts were tagged and recorded with GPS before being documented using digital still photography and high-definition video. Artifact locations were then recorded using baseline trilateration between permanently installed datum points (Van Tilburg 2005:14). The archaeological data recorded during these two field seasons yielded a site plan marking all visible artifacts and topographic features (Figure 4.4) and created a descriptive artifact database (Van Tilburg 2006).

In 2008, the *Parker* site was revisited by a team from PMNM who photographed exposed artifacts and recovered the ship's bell. Collected under an Archaeological Resources Protection Act permit and a Special Use permit issued by US Fish and Wildlife Service and reviewed by the State Historic Preservation Division, the bell could potentially provide the identity of the wreck (e.g., if the ship's name was marked on it) and could later be used as an interpretive device. It was photographed in situ, then sketched, tagged, and stored appropriately before being conserved. Although no name appeared upon completion of conservation treatments, the artifact was later included in a permanent exhibit on shipwrecks within the monument at PMNM's Mokupāpapa Discovery Center in Hilo (Fox 2010:8).

The *Parker* site was among several shipwreck sites inspected at Kure Atoll by both the 2010 and 2012 PMNM Maritime Heritage Cruises. The primary objective for these visits was to monitor any impacts occurring over time and document relevant cultural materials. Each of the condition assessments indicated no noticeable impacts, and, despite the rough weather experienced at Kure Atoll each winter, the site is considered stable. A secondary objective entailed inspecting the artifact assemblage to classify the types or functions of

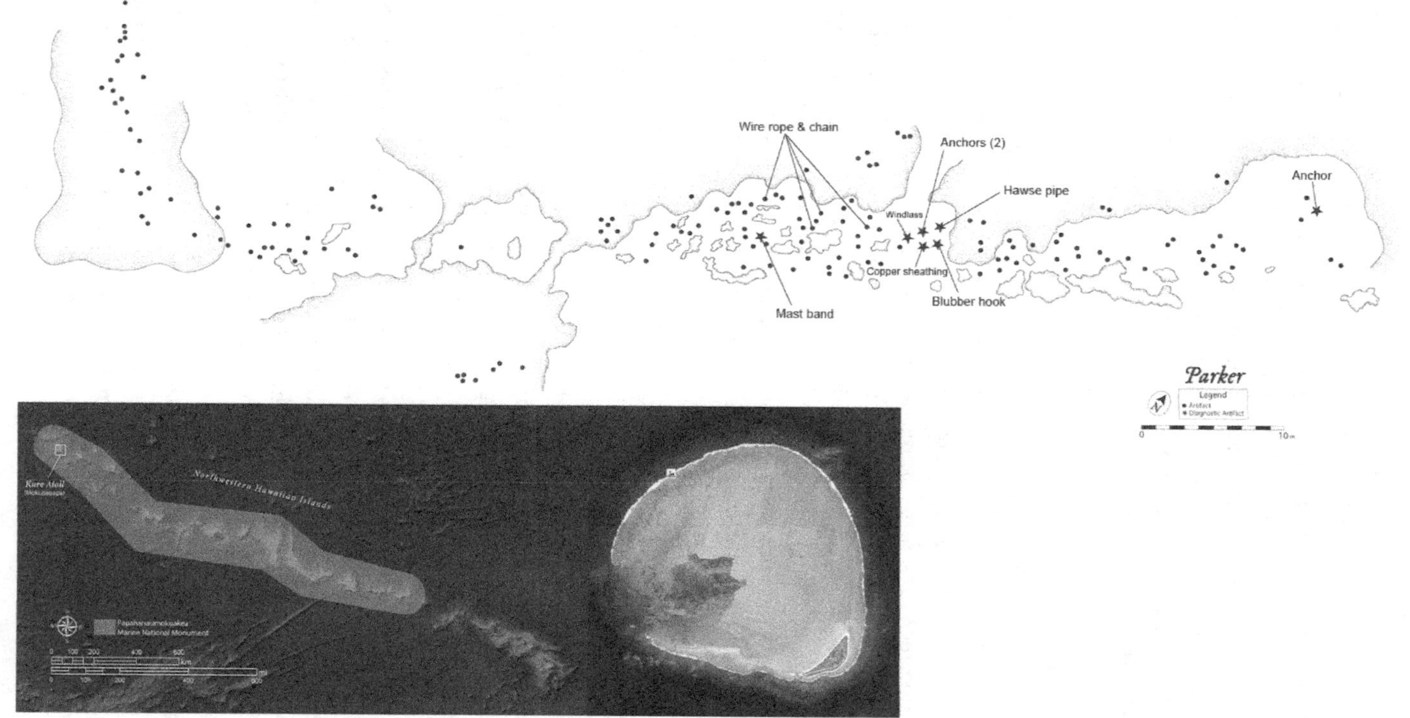

Figure 4.4. Site plan of the *Parker* shipwreck at Kure Atoll. (Courtesy of Jeremy Borrelli.)

the many previously unidentified objects. Prior to each of these expeditions, studies of material culture specific to pelagic whaling and the ships employed by the fishery were undertaken; this approach proved effective and numerous artifacts were positively identified. During each visit, archaeological photographs were taken of each artifact (or a representative sample where numerous artifacts of the same type exist). Additionally in 2010, marine remote sensing was again undertaken outside the reef to further refine the data acquired in 2003. Although rough sea conditions prevented side-scan sonar data collection, the 2010 magnetometer survey identified far fewer magnetic anomalies than those detected in the 2003 survey (SEARCH 2010:71–79).

Site Description

The wreck site of the whaleship *Parker* rests in the northwestern portion of Kure Atoll. The site is characterized by a scatter of artifacts stretching over an area approximately 400 m in length by 30 m in width with components existing on the outer reef, back reef, and inside the lagoon. Depths at the site range from between 2 to 5 m, and visibility generally ranges from 5 to 20 m. Most site features are in a low energy area inside the lagoon on a bottom consisting of patch coral reef, coralline substrate, and rubble and sand areas (Van Tilburg 2006). Coral growth at the site is highly variable; cover on the surface reef is minimal, whereas that on the back reef is dense with a wide array of coral species (Dana 1971:83). Coral encrustation on the cultural material in the areas inside the lagoon has been described as abundant, with cauliflower coral (*Pocillopora ligulata*) and lobe coral (*Porites lobata*) frequently encountered (Dana 1971:83; Van Tilburg 2003).

The scattered remains of the whaleship *Parker* stretch in a line along a scour channel from the ship's point of impact with the reef into the lagoon (Van Tilburg 2003:33). Though the largest concentration of artifacts lies in the sheltered confines of the lagoon, some artifacts were located on the surface reef and in a channel leading to the back reef area. Although diving surveys in these areas are difficult due to dangerous swells, they have identified fasteners, bricks, iron straps, and broken try-pot shards (Van Tilburg 2003:33, 2006). No wooden structural elements of the ship were noted on the surface reef or back reef areas; between the effects of the violent storm that caused the ship to quickly break apart and the dynamic environment, it is unlikely that any such remains exist.

The area between the back reef and the main concentration of artifacts at the *Parker* site consists of shallow patch reefs and sand pockets. Though the shallow depths in much of this section make surveying difficult, portions of it were explored in 2003, 2006, and 2010 to further delineate the extent of the site. These surveys resulted in the location of numerous diagnostic artifacts including a ship's bell, a possible gudgeon, four cupreous rudder pintles, cupreous fasteners, copper sheathing, and the bottom half of a stoneware container, as well as numerous metal objects of unidentifiable function. All these objects were presumably deposited as the ship broke apart after being pushed over the reef. Since these remains indicate that at least a portion of the ship's stern came to rest there, it is likely that other remains are buried in the area's many sand pockets.

The main section of the *Parker* wreck site consists of a concentration of scattered artifacts measuring approximately 75 m by 15 m. Beginning roughly 300 m south-southeast of the proposed point at which the ship impacted the surface reef, this section of the site rests largely in a sand pocket between patch reefs. Artifacts identified include anchors, scattered chain, copper fasteners of varying size, iron strapping, mast hoops, rigging elements, wire rope, deck machinery, copper sheathing, brick, and hawsepipes (Van Tilburg 2002a:54). Analysis of these cultural materials indicates an association with the bow of a large nineteenth-century wooden sailing ship (Van Tilburg 2002a:53–54). A small number of the copper driftpins found in this area were found to have very small portions of wood adhering them. Though wood sample analysis can be extremely helpful in determining the origin of a ship's construction, the conditions of these particular remains were too degraded and insubstantial in size for such testing to be successful.

Evidence for Identification as *Parker*

Diagnostic materials suggest that the wreck is that of an early to mid-nineteenth-century whaleship. These include numerous pieces of rigging and other materials associated with a wooden, sail-powered vessel, as well as multiple bricks and try-pot shards found at the site. A review of the historical record of ships known to have wrecked at Kure Atoll during this period revealed two candidates: the British whaleship *Gledstanes* wrecked in 1837 and the American whaler *Parker* wrecked in 1842. Since American and British whaling vessels of the period were outfitted with similar equipment and the amount of time between the losses of these two ships was relatively short, positive identification of the wreck proved difficult. Careful analysis of the physical remains at the site and of historical records, however, coupled with continued archaeological research at the atoll, ultimately led to the determination that the wreck was that of the whaleship *Parker*.

The main indicator for identification resulted from comparisons of the physical evidence found at the site with the descriptions of the wrecking events of each of the whaleships lost at Kure Atoll. Site analysis indicates that the ship impacted the northern part of the reef at a bearing of 135 degrees (magnetic) with enough force to be driven over it and into the lagoon. As the ship was violently pushed over by the heavy swells, it created a scour channel in the reef crest and some deck features were lost overboard (Van Tilburg 2006, 2010:312). Once inside the shallow lagoon, swells continued to wreak havoc on the vessel, breaking it apart and scattering remains along a trail before depositing the bow section nearly 300 m from the point of initial impact. Using this interpretation as a model, data from survivor's accounts of *Gledstanes* and *Parker* was scrutinized to determine the best match.

The British whaleship *Gledstanes* wrecked at Kure Atoll on the night of June 9, 1837. According to an account provided by the ship's master John Richard Brown, the 428-ton London whaler was cruising for whales when Ocean Island was sighted approximately 20 km, southwest by south. Thinking that the strong northern current they experienced throughout the day would help to keep them clear of danger, the captain ordered the course changed to southwest and the sails shortened to reduce speed. These efforts proved futile, and *Gledstanes* struck the reef near midnight, causing the crew to immediately launch three boats and abandon ship (Sandwich Island Gazette 1837). Based on Captain Brown's description of the wrecking event and an accompanying map, it appears that *Gledstanes* contacted the east side of the reef since Ocean Island was first sighted at a bearing of roughly 200 degrees (Hawaiian Spectator 1838). Further evidence of the location of the *Gledstanes* site was provided by an officer of USS *Saginaw*, a USN gunboat lost at Kure Atoll on October 29, 1870. The survivors of that wreck noted seeing some wooden remains of *Gledstanes* washed up on the beach and others emergent on the reef (Read 1912:12; Van Tilburg 2010:229). The location of the wreck of *Gledstanes* was included on a map of the atoll (Figure 4.5) made by Lieutenant Commander Montgomery Sicard; it is positioned to the north of the spot at which USS *Saginaw* hit the reef (Read 1912:12). Considering that the location indicated on both maps is almost exact, it is evident that the wreck in the north-northeastern section of Kure Atoll was not that of *Gledstanes*.

The details of the loss of the whaleship *Parker* provided by its shipkeeper offer a much more plausible identification for the wreck in the north-northeastern section of Kure Atoll. From the description presented in the survivor's account, the sighting of Ocean Island to the southeast suggests that the ship was sailing to the north and west of the atoll. With no change in course indicated, and with heavy winds blowing from the north to northeast, it is likely that the ship was driven on a direct trajectory toward the reef. While no maps marking the location of the wreck of *Parker* are known, Quinn's account states that the ship "struck the reef about eight miles N.N.W. from the center of the Ocean Island" (Temperance Advocate and Seamen's Friend 1843),

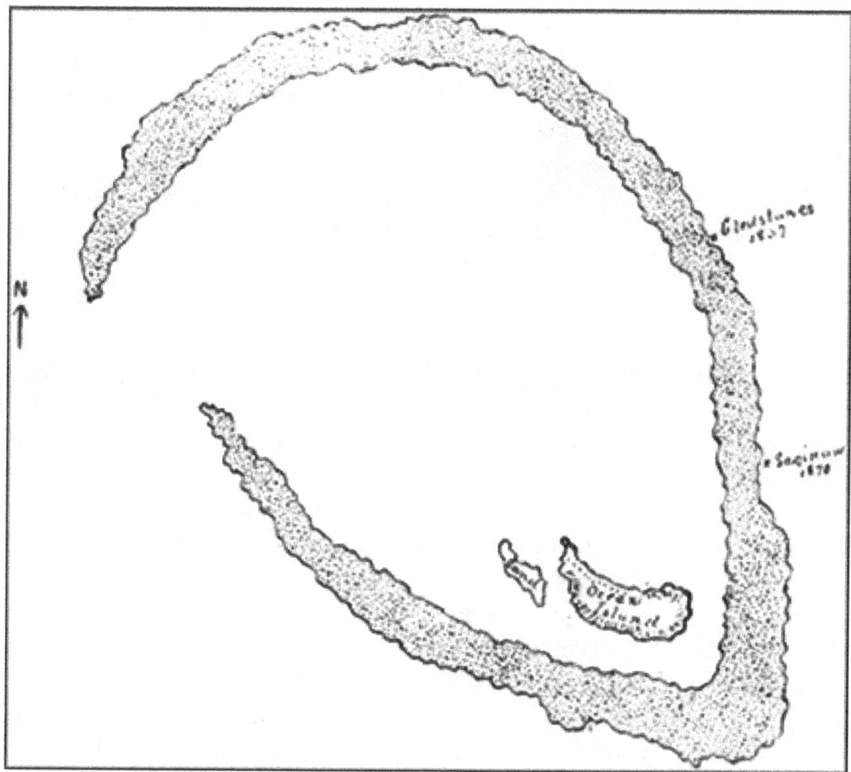

Figure 4.5. Illustration of Ocean Island and reef by Lieutenant Sicard indicating the locations of the wrecks of *Gledstanes* (1837) and USS *Saginaw* (1870). (From Read 1912)

which positions the point of impact as being along the north-northeastern part of the reef. Though this reported distance is slightly farther than the actual distance of approximately 9 km from the center of the island to the archaeological site, it is also the farthest distance from the center of the island to any point on the reef. When compared with data gleaned at the wreck site, these observations closely match the physical evidence and indicate that the wreck in the north-northeastern section of Kure Atoll is almost certainly that of *Parker*.

One other significant piece of evidence supporting the identity of the remains as belonging to the whaleship *Parker* came with the location of the wreck of *Gledstanes*. Though the exposed location of its loss made accessing the spot marked on the hand-drawn map difficult, a period of calm weather in 2008 allowed for surveys close to the reef. These resulted in the identification of the remains of a whaleship with diagnostic artifacts including anchors, iron tryworks knees, a partial try-pot, chain, bricks, and iron ballast. While most of these artifacts are indicative of any wooden sail-powered whaleship of the period, the presence of pig iron bars was key to positive identification. British sailing vessels of the eighteenth and nineteenth centuries carried iron ballast bars (often called pigs) measuring approximately 90 cm by 15 cm in their lower holds to help the ship sail upright (Roberts 1992:55; Tuttle et al. 2010:43). Since most of the pigs documented on the site at Kure Atoll closely matched those dimensions, it was highly likely the wreck is that of the British whaleship *Gledstanes*.

Conclusion

Many dangers awaited ships passing through the NWHI. The unpredictably of the open ocean environment in that region of the Pacific and its uncharted yet treacherous reefs acted as a trap for ships. From the earliest reports of immense sperm whale populations existing at the northern end of the archipelago through the end of industry's golden age, whaleships were the most common vessel traffic into the region. In fact, aside from the 10 whalers lost there between 1822 and 1867, only one other wreck is known to have occurred: the merchant ship *Huntress* ran aground in 1852 on a voyage between San Francisco and Hong Kong (Polynesian 1852). The designation of the NWHI as a marine protected area and the commitment of PMNM to recording, studying, and managing maritime heritage have resulted in identification of half of the sites of historically known whaleship losses. The exemplary research undertaken at these remote archaeological sites provides new insight into early to mid-nineteenth century pelagic whaling. The results of two of the most thoroughly documented and studied wrecks, those of *Two Brothers* and *Parker*, are considered in the following chapters.

5

The American Whaling System during the Golden Age

The basic structure for the American system of whaling developed in the mid-eighteenth century. It involved not only the specific industrial processes and specialized technologies used at sea to capture whales and extract oil but also the human effort that effectively drove the process. This aspect of the system included a division of labor that prescribed strictly defined roles, a wage structure that rewarded hard work and successful hunts, an operational schedule that maximized profit potential though vessel and equipment maintenance, and risk management overseen by a hierarchical labor structure. All aspects of the system combined to produce a specific workplace that only existed aboard whaling vessels of the period. This in turn produced a distinctive assemblage of material cultural remains and archaeological signatures. This chapter begins with an explanation of pelagic whaling as a three-part system before describing in detail the stages that bookended a voyage and enabled ships to become the seaborne industrial workplaces that is discussed in the following chapter. It also illustrates the ways in which those stages are reflected in the distinctive assemblage of material cultural remains and archaeological signatures of wrecks of the whaleships *Two Brothers* and *Parker*.

Pelagic Whaling as a System

To understand how the remains of American whaleships are interpreted as maritime industrial workplaces, it is first necessary to briefly explain the larger system within which the fishery operated. By the turn of the nineteenth century, pelagic whaling operations were mostly standardized and determined by stages of a voyage. For this discussion, the pelagic whaling system is divided into three main stages: the precruise stage, in which investors agreed to finance a whaling cruise and all necessary preparations for it were made; the cruise operations stage, which included all relevant industrial operations of a vessel from its departure to its return; and the postcruise stage, which involved releasing crews, settling accounts, cleaning the vessel, and preparing it for the next voyage. Each of the three interrelated stages represents a culmination of operations and processes, and none was more critical than the others.

Because of their importance to the overall success of pelagic whaling, each stage is considered, to some extent, as part of the significant historical interpretations of the fishery. In some cases, particular aspects of each stage have been thoroughly examined to provide insight into their role in the success or failure of a whaling voyage (see for example Davis et al. [1987]; Davis et al. [1997]; Finney [2010]; Hohman [1928]; Jenkins [1921]; Tower [1907]). It is the cruise operations stage of the system, however, that is most described in the multitude of works dealing with historic whaling. Almost every history of the industry contains sketches—some sensational, others factual—describing the call of "thar she blows" from the masthead, of whalers pursuing their prey, of the thrills and perils associated with the chase, and sometimes of the filthy, disagreeable task of boiling blubber into oil. Undoubtedly, those industrial processes were instrumental to the financial success of a whaling cruise, and many firsthand accounts accurately portray them in that light. But when viewed as a part of the larger picture of life at sea on an early nineteenth-century ship, they were only sporadic bursts of activity that interrupted the otherwise mundane—but just as crucial—aspects of industrial life onboard.

Between 1820 and 1840, the average length of an American whaling voyage extended from roughly two years to four, and during most of that time, crews were occupied with routine aspects of the whaling life. Conducted under the supervision of officers, these strictly regimented activities were interrelated, and adherence to the daily structure meant that the ships were ready for action as soon as they sighted whales. The decisions and preparations made during the precruise stage were essential to the safety and success of a voyage, and the accounting activities of the postcruise stage officially ended the voyage and freed the crews from their voluntary confinement.

Precruise Stage

The precruise stage of the pelagic whaling system was perhaps the most important in the success of a venture due to the remoteness of the field of operations and the requirement for careful planning. By the early nineteenth century, most whaleships leaving American ports were bound for the largely uncharted expanse of the Pacific Ocean. As such, it was necessary that the ships carried onboard everything they could possibly need to undertake industrial operations—including stores of food, fresh water, and medical supplies that might need to last several years, since there was no guarantee of places to reprovision. Thus, from the initial steps of securing sufficient financial support and the appointment of experienced managers to oversee all aspects of outfitting and operations, the decisions made at this stage required the utmost consideration and directly affected the financial success of a voyage as well as the safety of the crew.

Investors and Agents

Long before a whaleship ever set sail, an investor or group of investors first agreed to provide the capital for the voyage. In the eighteenth century, when Nantucket ships dominated the industry, investment in cruises was generally a group venture by members of the community who together accepted all risks involved, as much as they did the potential for profit. Community members, officers and crew, vessel owners, rope makers, blacksmiths, bakers, and many other merchants responsible for provisioning a ship all contributed with the knowledge that their payment depended on the success of the voyage (Fairburn 1945:257). As the center of activity shifted from Nantucket Island to coastal towns on the mainland, the system of ownership in whaling cruises changed to favor more capitalist approaches. By the early nineteenth century, ownership was a more consolidated affair, and the number of investors decreased to the extent that voyages were more likely to be coordinated by agents from whaling firms who invested only with close business associates and/or small family groups (Finney 2010:74).

Regardless of the number of investors, the business of whaling began and ended with the agents (Finney 2010:75). Agents were entrepreneurs who took responsibility for managing the entire whaling voyage, from the initial investment to the finalization of financial and legal matters. After 1819, it was standard that agents were also part owners in the venture; this meant that they took on great financial responsibility and risk, which consequently helped to instill confidence among investors (Davis et al. 1997:400). Once they procured the capital, the agents acted on behalf of the owners and made all necessary decisions and preparations for the cruise. It is important to note that although the steps in planning for a prospective voyage were not always made in a particular order, each decision affected choices made about other matters relevant to the cruise (Davis et al. 1997:214).

Engaging a Master

One of the most important decisions an agent made was procuring the services of an experienced whaling master to oversee operations while the ship was at sea. Though these individuals were addressed as "captain" while onboard (Finney 2010:2), the title of master is perhaps more apt given the responsibilities pursuant to the position. Commanding a whaling vessel demanded a range of talents and tremendous responsibility (Vickers 1985:283), and as such, the master "was given full responsibility and supreme authority over every detail of the cruise and every crew member[,] . . . and although accountable for his conduct to the courts and owners ashore, he was the most unbounded of autocrats at sea" (Hohman 1928:119). As with the agent, masters also often invested in whaling voyages, which meant that management of provisions and efficiency of whaling operations were even more important to them. Their responsibilities included determining hunting strategies and understanding seasonal movements, conducting all ship's business, and doctoring the sick, as well as governing the officers and crew and, most importantly, returning to port with a full hold (Shapiro 1959:22). In addition, whaling masters also required a high degree of literacy and a greater proficiency in sailing and navigation than the captains of other types of contemporary vessels. As with all sailing ships of the period, whaleships were among the most complex machines in use, and operating them required a great deal of knowledge and skill (Sievers 2009:402; Vickers and Walsh 2005:88). In contrast to contemporary merchant ships, which generally traveled on well-established routes between ports, for the most part, early nineteenth-century whaleships operated in mainly uncharted waters, necessitating greater skill for safe passage (Finney 2010:2).

Grounds to Be Hunted

Once the agent and master agreed to terms of the contract, the next decision was the determination of the primary grounds to be hunted. "Grounds" were areas where specific species of whales congregated, fed, mated, or calved and were thus more likely to be found (Finney 2010:46; Leavitt 1973:125; Morton 1982:85). These areas were not contained within rigid physical boundaries but instead were general parts of a particular ocean; thus, this knowledge allowed whalers to plan for seasonal hunting patterns (Davis et al. 1997:108). Since opportunistic whaling occurred as soon as a ship left its home port, the choice of a primary whaling ground affected the route taken (e.g., passing through other known grounds), necessary modifications to the vessel based on expected conditions (e.g., additional hull reinforcement for specific environments), and the projected length of the voyage (e.g., amount of outfitting and provisions required).

Noticeable depletions in whale populations on the Atlantic grounds led whalers to round the continental capes and begin fishing in the Pacific Ocean by 1789 (Kugler 1980:22) and the Indian Ocean by 1791 (Wray and Martin 1979:213). The waters of the Pacific proved to be well stocked, and by 1819, the locations of several grounds in the south and central parts of the ocean were commonly known. When hunting on those grounds, whaleships pushed farther offshore hoping to expand boundaries or find new areas. Over time, this practice helped to develop a better understanding of the seasonal migration patterns of whales in the Pacific. Thus, once the whaling agents and masters agreed on a primary ground, the master drew upon personal experience as well as reports from others to determine the movements of the ship while at sea (Kugler 1980:22). A review of Alexander Starbuck's (1878) list of "Returns of Whaling-Vessels, Sailing from American Ports" indicates immense variability in the detail offered by agents regarding this decision. Though some ships are simply listed as having sailed for the "Pacific" or a general area of it (e.g., north or south), others are more specific and provide the name of the actual ground to be fished (Starbuck 1878:166–167).

Vessels and Rigs

As mentioned, the decision to hunt on a particular whaling ground had a direct effect on subsequent decisions regarding the physical aspects of voyage preparation. Primary among those decisions was the choice of a vessel that was appropriate for the cruise. The ships employed in sperm whaling in the early to mid-nineteenth century were the most important factor of the whaling business (Hegarty 1964:18). Not only did the ship provide the vehicle for transportation to and from the hunting grounds and the platform for onboard industrial operations, it also acted as a home away from home, offering accommodation and other spaces that officers and crew shared over the long voyages. Thus, the choice of a suitable vessel and its continual maintenance were tantamount to the success of a voyage.

As the length of voyages grew longer and as the seas hunted more remote, whaleships grew in size and complexity of their outfitting. By the 1820s, ships often spent three or more years at sea, where they experienced climates that ranged from blistering heat to freezing cold. Over the course of a voyage, a ship's timbers were slowly weakened not only by punishing environmental elements they routinely encountered but also by the silent, destructive infestation of wood-boring organisms such as toredo worms (Gibson and Whitehead 1993:134). As such, all these expected conditions and potential dangers were considered when a ship was chosen for a whaling cruise.

The primary decision regarding the choice of a vessel was whether to reuse an existing hull or construct a new one. Though recycling hulls from the merchant service had long been practiced by whaling agents, the aftermath of the War of 1812 saw a dramatic increase in the numbers of those ships being employed in the fishery, which made finding them difficult. Thus, the 1820s and 1830s saw the advent of the purpose-built whaleship (Mawer 1999:253). The obvious advantages to new construction were the ability to incorporate the latest technical improvements (Davis et al. 1997:216) and to achieve optimal cargo capacity. Although new hull construction became more common and preferred, it was nonetheless expensive. Nantucket historian and whaleship owner Obed Macy lists the cost of new whaleships built in Nantucket in the 1820s as averaging $22,000. He explains that such costs were necessary to produce vessels "fit for the arduous and protracted voyages they are destined to perform" (Macy 1835:221). Construction expenses included fees and shipping costs for raw materials such as timber, copper fasteners and sheathing, cables and lines, anchors and chains, and ballast. Other costs included labor for the tasks performed by carpenters, caulkers, blacksmiths, painters, riggers, and sailmakers (Hohman 1928:324; Macy 1835:221; Schultz 1967). These costs continued to grow with the numbers of ships entering the fishery, and by 1841, a newly constructed New Bedford whaleship required an outlay of over $31,000 (Hohman 1928:324).

Because of the high costs, many agents opted to recycle any existing hulls that could be found. Some of the recycled vessels were used in the merchant service and required some reconfiguration; others had been employed in whaling for a short time and needed only relatively minor refitting; and still others were whalers beyond retirement age but that could be acquired cheaply due to the extensive work needed to make them seaworthy (Davis and Gallman 1994:58; Davis et al. 1997:216). Because so many different types of vessels were converted for use as whalers, no one hull type is considered typical. Whaleship historian Joseph T. Higgins (1927:5) explains the variations in hull form by stating, "some were deep and some shallow, some blunt and some sharp, some had flat transoms and others had shallow ones that were curved." Despite this range, the type most favored for reuse was the transatlantic packet ship, which was intended for carrying passengers, mail, and cargo. Since speed was not a requirement and the desired characteristics of a whaling ship were that they be extremely seaworthy, small, and burdensome with a full buoyant bow, the hulls of packets proved ideal, so many were converted for whaling (Fairburn 1945:1627).

If the decision was made to redeploy an existing whaling vessel, a suitable candidate could often be found tied alongside the wharves of New England's ports awaiting return to service (Littlefield 1906:4). Interestingly, a whaleship's age was not always a concern. Despite the harshness of the conditions to which they were subjected, whaleships often had lengthy working careers—some as long as 90 years (Church 1938:19–20; Davis et al. 1987:48). While the diligence of their masters regarding maintenance was undoubtedly the main contributor to this longevity, some speculate that the preservative qualities of whale oil that leaked from the casks and saturated the wood also played a role (New York Times 1975). As maritime historian William Fairburn suggests, "the whaling service seems to have operated actually to preserve the timbers and planking of the wooden vessels steadily engaged therein, as the oil impregnated the structure and prevented rot and general deterioration" (Fairburn 1945:1007).

Regardless of whether the hull was reused or newly constructed, its size was an important factor. Size was recorded in tonnage, which until after 1865 was a simple measurement based on overall length, breadth, and depth of hold (Davis et al. 1997:215; Morrison 1921:14). Factors affecting size included the capacity of holds for carrying oil, the number of officers and crew required, and voyage duration (Fairburn 1945:1022). In the rush to reestablish Pacific whaling immediately following the end of the War of 1812, much of the fleet was comprised of a variety of available vessels, many of which proved to be too small. Peacetime, however, allowed the Pacific whale fisheries to flourish and with this came capital increases; between 1820 and 1840, vessel size steadily increased from roughly 280 to around 400 tons (Spence 1980:99). Thus, an average whaleship of this period was around 350 tons (Davis et al. 1987:9) and measured approximately 33.5 m in length, 8.2 m in beam, and 4.2 m in depth (Leavitt 1973:13).

Closely aligned with the choice of vessel size was the rig composition (Davis et al. 1997:216). Though pelagic whaling vessels were generically referred to as "whaleships," several different rig configurations were employed including sloops, schooners, brigs, ships, and barks (Davis et al. 1987:23). Prior to entering the Pacific Ocean, whalers commonly used schooners and brigs for Atlantic voyages, which generally lasted less than a year. By 1800, however, extended voyage lengths required more substantial platforms and increased cargo space, which resulted in the full rigged ship becoming the most common rig used for vessels destined for the Pacific grounds (Kugler 1980:21). Vessels configured as ships utilized three masts, all of which carried multiple square sails hung from yards attached perpendicular to the mast, as well as triangular jib sails rigged to a large bowsprit (Paasch 1885:1). With the larger spread of sails came increased speed; and because they could be employed on vessels of greater carrying capacity, the ship rig became the most popular with agents and masters in the early decades of the nineteenth century.

By the 1850s, the bark became the dominant rig in the American fleet due to changes in hunting grounds, which required more maneuverability (Kugler 1980:21). Having the advantage of greater maneuverability, the bark rig proved itself among the dangerous ice floes encountered in the Arctic and eventually replaced the ship rig altogether (Davis et al. 1997:270). Whaling barks were three-masted sailing vessels that carried square sails on the fore and main masts, fore-and-aft sails on the mizzen, and jib sails attached to a large bowsprit (Leavitt 1973:122). Though this rig had previously been limited to ships of less than 100 tons, by the 1830s improvements in geared winches, iron-strapped blocks, geared steering, better mast and spar ironwork, and lighter canvas for sails all combined to enable their use onboard whaleships (Chapelle 1967:279; Davis et al. 1987:26; Davis et al. 1997:270).

The ships *Two Brothers* and *Parker* are indicative of trends in their period of use. In the first instance, information pertaining to the two known voyages (1818–1821 and 1821–1823) undertaken by the Nantucket whaleship *Two Brothers* indicates that it was a 217-ton vessel and rigged as a ship (Starbuck 1878:226–237). Because the original construction data for this ship is unavailable, it is unknown if it was purpose-built for whaling. Its relatively light tonnage, however, may be seen as a solid indicator of the reuse of an older hull since, according to Macy (1835:221), whaleships built at Nantucket immediately following the War of 1812 averaged 300 tons. The New Bedford whaler *Parker*, however, was built specifically for pelagic whaling. Constructed in 1831, customs records indicate that it was a 406-ton, ship-rigged vessel that measured 34 m in length by 8.5 m in beam and 4.2 m in depth (WPA 1940:248). These registered physical dimensions closely match those for an average whaleship of the 1820–1840 period, though the tonnage is at the upper end of the weight range (Davis et al. 1987:9; Leavitt 1973:13; Spence 1980:99).

Crews

An agent's decisions regarding the grounds to be hunted, the vessel's size, and its rig also had implications for the number of crew required (Davis et al. 1997:228). Though many personnel were certainly needed to handle the numerous, massive sails and rigging of any ship or bark, the main determining factor in the size of a whaleship's crew was the number of whaleboats it carried (Tower 1907:89). Although larger crew numbers meant increased operating costs for the owners, the more whaleboats that were hunting directly increased profit potential. Since the effectiveness of the whaleboats was crucial to success, over time a standard crew structure developed; each boat required a boat header, boatsteerer, and four or five average sailors to operate the oars (Tower 1907:89). Thus, the personnel of an early to mid-nineteenth-century whaleship carrying four boats consisted of the captain, four mates (boat headers), four boatsteerers, a cooper, blacksmith (though these duties were often performed by the cooper), carpenter, cook, steward, cabin boy, and 16 to 18 foremast hands—roughly 32 crew members in total (Brown 1887:221; WPA 1938:13). Because the number of crew required was greater than those of merchant vessels of similar size and rig, their presence onboard became a distinguishing feature of nineteenth-century whaleships (Brown 1887:234).

The size and quality of the crews required to operate American whaleships changed over time. In the earliest period of pelagic whaling, the typical Nantucket whaler was a small sloop of fewer than 80 tons and crewed by 13 hands who had a gone to sea as a calling or a profession (Hohman 1928:51; Sievers 2009:402). These early whale fishers followed a long-established tradition on Nantucket that saw them first join the crew of a whaleship in their early teens (often following the lead of a family member) and embracing it as a career path by learning the skills required of each role before eventually taking command of a ship in their early twenties (Sievers 2009:402). These crews were considered hardy and expert mariners, and the American whaling fishery was revered as a nursery for the merchant marine (Fairburn 1945:172; Olmsted 1841:120; Stackpole 1953:317). The shift in industrial dominance from Nantucket to New Bedford, however, saw this old system decline. The explosion in the number of vessels engaged in the fishery, the increase in their size, and the changes in rig saw the need for a greater number of crew. By 1825, vessel size was closer to 300 tons and crews numbered 25 (Hohman 1928:51). The increase in crew numbers inadvertently resulted in a noticeable decrease in the quality of hands recruited, which in turn had a direct effect on the corporate knowledge and skills of the officers. Historian William Fairburn (1945:1037) suggests that this was the result of exploitation and abuse on the part of increasingly unscrupulous owners and that "young Americans with enterprise and enthusiasm gradually became disillusioned and increasingly disgusted with whaling in the 1830s and 1840s, and when the supply of American youth failed in the forecastle, the quality of men for the cabin was lessened."

In the early decades of the 1800s, whaling agents personally engaged in recruiting and often relied on the advice of the ship's master to find experienced officers and crew. But by the 1830s, they depended more on outfitters, or shipping agents, for recruiting (Dolin 2007:221; Hohman 1928:70). Often referred to as "land sharks" (Brown 1887:289–290), these companies or individuals assisted with outfitting the crew and for this service received a small fee from the agent and the exclusive rights to equip recruits with essential clothing and gear, for which they charged exorbitant prices (Dulles 1933:94; Finney 2010:233; Verrill 1916:88). If using a shipping agent, the whaling agent provided the number of crew required and roles to be filled; this allowed the whaling agent to return to the task of managing the overhaul of the vessel in preparation for the cruise. Shipping agents devised effective methods to target would-be whalers, including posting attractive advertisements in coastal towns and regional newspapers that offered advances on pay and painted "glorious verbal pictures" of the easy life at sea, the huge profits to be made, and exotic locations to be visited (Dolin 2007:221; Dulles 1933:94; Verrill 1916:88). And there was no shortage of interested parties ranging from experienced whalers to neophytes known as greenies or green hands (WPA 1938:28).

By the 1830s, crews were increasingly drawn from all parts of the world and social strata; sons of the famous New England whaling families hoping to carry on a tradition could easily find themselves working alongside adventurous youngsters from inland farms, free Black seafarers, mariners from all nations of Europe, or even criminals (Bolster 1990; Hohman 1926:657). Experienced Native American whalers and Indigenous islanders from places like the Azores, Cape Verde—and by the 1820s, Hawaii and other Pacific islands—were excellent crew members on whaleships largely because of their reputations as skilled mariners (Hohman 1926:657; Lebo 2007:16). As a rule, labor onboard nineteenth-century whaleships was strictly a male activity (Norling 1996:71); though it was not uncommon for wives and families of captains to accompany them on a voyage, female involvement in actual whaling operations was rare (Verrill 1916:278–282).

American whalers were not paid a monthly wage, as were other merchant sailors of the time; instead they were offered a predetermined fractional share of the revenue (less costs) generated from the cruise (Creighton 1990:543; Davis et al. 1997:154; McConnell and Price 2006:295). Known as the "lay" system, this method of payment had long been used in the New England fisheries and was effectively a partnership between owners and crew members in which risks were shared and hard work was financially rewarded (Fairburn 1945:983; McConnell and Price 2006:296). Historian John R. Spears describes the effects of this "no oil, no pay" system on the whalers by stating that it "sharpened the eyes of the lookout, gave strength to the arm of the men at the oars, and cooled the nerves of the man who thrust the lance under the shoulder blade of the whale" (Spears 1910:201). The lays offered to mariners who signed on to a whaling voyage were relative to the amount of experience and the responsibility ascribed. The hierarchical structure of a whaleship's labor meant that the range of lays directly reflected the position in the chain of command or the importance of their skill. The lays offered to the captains and crew represented a percentage of the net profit; for instance, a 1/12 lay meant that the recipient received one barrel in every twelve (WPA 1938:18). In general, officers, boatsteerers, and highly skilled coopers might expect "short lays" ranging from 1/8 to 1/100; experienced mariners, stewards, cooks, and blacksmiths received shares from 1/100 to 1/160; while the greenies and cabin boys had to be content with "long lays" of between 1/160 and 1/250 (Grant 1932:36; Hohman 1926:645).

Through the contracts they signed, the crew also agreed to have fees deducted for outfitting, insurance, pilotage, wharfage, and cooperage, all of which affected their final earnings (Dodge 1882:7; Hohman 1926:645). Though the lay system was designed to provide incentive for whalers, there was no guarantee that a cruise would be profitable. The main factors that affected profitability included trouble finding prey and taking enough whales to make the cruise financially successful, debts incurred through cash advances from the captain or by purchasing goods out of the highly overpriced "slop-chest" or "skipper's store," and unexpected fluctuations in oil prices resulting from voyages that were longer than anticipated (Fairburn 1945:1016; McConnell and Price 2006:296; Morrison 1921:321). Because of these and other factors, quite often the hands on a whaleship holding long lays were lucky if they broke even at the end of a long and dangerous voyage.

Outfitting: Readying the Ship

Once an agent and master settled on a vessel and rig for a particular voyage and determined the number of crew needed, the next step in cruise preparation was to outfit the vessel. Although the distinction is sometimes blurred, outfitting can be seen as having two phases: readying the hull, masts, rigging, and sails of a vessel for the cruise; and stocking the ship with supplies necessary to sustain the long voyage (Davis et al. 1997:214). The first phase involved ensuring that the chosen vessel was physically equipped for the conditions expected. When a newly constructed hull was to be employed, the amount of work required for this phase was limited. When using a recycled vessel, however, the need to thoroughly inspect the hull and repair all weak or deteriorated components was of the utmost importance.

As was customary of the period, at the conclusion of each voyage the hull of a whaleship was completely emptied, and the rigging and spars were stripped (Littlefield 1906:4). Doing so lightened the ship and allowed for easier inspection through the process of careening. Careening involved hauling the vessel over on one side to expose the portion below the waterline and then replacing any rotten timbers, planking, and fasteners (Doane 1987:260). Once the shipbuilder was satisfied with the vessel's structural integrity on the exposed side, the exterior of the hull was prepared for the voyage. Pitch-pine boards 7.6 to 12.7 cm thick were attached to the outer face of the frames and then the seams between them were made watertight by thoroughly caulking with oakum and pitch (Hall 1884:25–26). To protect it from marine borers, the part of the hull that remained below the waterline was first coated with either 2 cm cedar planks (Spence 1980:101) or with a thin layer of cement or tar and "paper" (felt). A fresh layer of copper or "yellow metal" sheathing was then attached to the keel, rudder, stem, sternpost, and planking (Hall 1884:27; Ronnberg 1985:183). Any modifications to the interior of the hull were also made at that time. For instance, if a ship was intended to work in the Arctic, the bow would be reinforced with thick oak planks and an iron shoe fitted over the fore foot to protect against heavy blows from ice (Littlefield 1906:6). When hull preparation of the first side was completed, the ship was refloated and repositioned before the careening process was repeated for the other side.

As with most wooden ships of the early nineteenth century, aside from timber dowels (known as treenails), which were sometimes used to attach hull and ceiling planks (Hall 1884:26), all of the structural timbers and various finishing pieces were joined using metal fasteners. Fastenings ranged in size, type, and metallic composition depending on their planned usage. During this period, fasteners intended for use above the waterline were usually iron, while those installed below were made of metals that resisted corrosion such as copper, brass, bronze, or other composite metals (Burns 2003:56; Nantucket Historical Association Research Library [NHA] 1834). A review of documents relating to the construction and outfitting of the newly built whaleship *William Rotch* (1819) illustrates the many different types of fasteners used in its construction. Among those listed in accounting records were copper bolts, ring bolts, spikes, deck spikes, staples, nails, copper nails, pump nails, deck nails, mast nails, cut nails, brads, tacks, copper tacks, and screws (New Bedford Whaling Museum Research Library and Archives [NBWM] 1819). Though numerous, these fastener types are only a sample of the fittings used in whaleship construction and fitting out; as the nineteenth century progressed, the variations in size, shape, proportion, and manufacture of fasteners were almost endless (Higgins 1927:27–28).

As with fasteners, the sheathing attached to nineteenth-century whaleships for protection from wood-boring marine organisms was also typical of that used for other wooden ships of the period. Although the practice of covering hull planking with extra material can be traced back thousands of years, the different coatings used for this purpose were only minimally successful (McCarthy 2005; Staniforth 1985). Experiments with copper in the early to mid-eighteenth century, however, showed it to be the first truly effective barrier against fouling (McCarthy 2005:102–103; Stone 1993:23). The use of copper as a sheathing material not only resisted the attack of the shipworm (*Toredo navalis*) and the gribble (*Limnoria* species), the cleaner bottoms also achieved greater speeds, were more maneuverable, and required less time for repairs (Bingeman et al. 2000:222; Staniforth 1985:21). A major disadvantage in the early period of its use was the galvanic action that occurred due to contact with copper and a ship's ironwork (McCarthy 2005:103). Nevertheless, this issue was solved with the development of various copper-alloy fasteners that were much harder than pure copper and could be driven into hardwood (McCarthy 2005:105–107; Staniforth 1985:26).

Although there is some speculation that the Chinese, and then the Dutch, had employed copper sheathing on the hulls of vessels as early as the seventeenth century, it was the adoption of this technology by the Royal Navy that set the standard for its use (Bingeman et al. 2000:220; McCarthy 2005:102). In the 1770s, copper sheathing was attached to the hulls of some British merchant ships, and by the end of the eighteenth century, it was common for French and American craft to also be "coppered" (Stone 1993:23). In a relatively short time, this practice became standard for all oceangoing vessels, which led to an increase in demand and cost for copper. Although the copper used for sheathing ships was pure throughout the early part of the nineteenth century, experiments with the different alloys were undertaken. In 1832, a copper-zinc alloy known as yellow-metal or Muntz proved to be tougher, longer wearing, and less expensive than copper (Crothers 1997:330; McCarthy 2005:115–121; Ronnberg 1985:183; Stone 1993:23). These qualities were quickly realized, and by the 1850s, Muntz metal—so named after its inventor—replaced copper as the most widely used sheathing metal (Burns 2003:63). The increased use of copper-alloy sheathing for whaleships is evident in the growing number of advertisements in the *Whalemen's Shipping List* over the mid-nineteenth century, and by 1860, most whaleships were sheathed with it exclusively (Hegarty 1964:49; Ronnberg 1985:183).

Archaeological Evidence for Readying the Ship

Archaeological evidence of ship architecture is found at both the *Two Brothers* and *Parker* shipwreck sites. Although the wooden elements of each of these ships have long since vanished (Thomson 1997:124), their metal components are less susceptible to biodegradation or the harsh subtropical environment of PMNM. Thus, their architectural remains are mostly pieces and fragments of sheathing metal and the fastenings that either held timbers together or were kept onboard for repairs. In general, the context for these fasteners is distorted from the timbers that once held them; this is because the same conditions that worked to smash these ships and destroy their organic structures also tended to disperse the remaining fasteners and sheathing over large areas (McCarthy 1982:1).

Several fasteners were found among the remains of *Two Brothers*. These included bolts, spikes, nails, and tacks; all but one were manufactured using a copper alloy. Although recorded at both sections of the site, most fasteners were found in Section B. Several factors are thought to contribute to this disparity. The large area encompassed by Section A, its pocked reef top environment, the scattered nature of the remains located there, and the difficult sea conditions at the time of investigation, all made identifying objects as small as fasteners difficult. As a result, only three fasteners were identified in Section A: a portion of a 15-cm-long copper bolt and a 10-cm-long copper nail—both of indeterminate function but clearly intended for use below the waterline—and a partial iron bolt of indeterminate function that measures roughly 20 cm in length and 2.5 cm in diameter.

In Section B of the *Two Brothers* site, numerous copper-alloy spikes and tacks were found in pockets in the reef and on the seabed. The intended uses for these types are known; square-headed spikes of the size found, 13 to 16 cm in length, were often used for attaching planks to the hull below the waterline and for general fastening, while the 2.5-cm-long tacks were probably used to secure sheathing plates (Steffy 1994:289) or for repairing damaged or building new whaleboats when necessary. Most of the spikes showed signs of wear such as bending or breakage, which indicates that they were in use and damaged during the wrecking event. Though a small number of copper-alloy tacks were scattered throughout the southern part of Section B, most were found in two pockets in the reef. These concentrations of tacks most likely indicate that they were unused. Instead, it is probable that they were stored in wooden kegs and subsequently deposited on the reef as the ship's stern broke apart; as the kegs deteriorated, the tacks settled in place.

Unlike the fasteners recorded at *Two Brothers*, most of those identified at the *Parker* site were used for securing structural timbers to one another. The majority of these were copper or copper-alloy drift bolts, which are thick, round fasteners commonly used for attaching major timber pieces such as the frames, keel, and keelson (Stone 1993:34–35). Often simply referred to as drifts, these bolts were cut from a rod to a specific length and then tightly driven into timbers via pre-augured holes. To produce an extra-strong joint and to prevent loosening, one end of the bolt was generally driven thoroughly down over a washer called a clench ring, while the other end was bent over for added strength (Stone 1993:34–35). Drift bolt length depended on the width of the timbers to be joined; those recorded at the *Parker* site ranged from roughly 30 to 80 cm. According to whaleship historian Reginald Hegarty (1964:27), whaleship keels were "built up of two lines of hewn oak logs 14 to 16 inches [35.5 to 40.6 cm] square, placed one upon the other and securely bolted together." Thus, the length of the larger bolts found at the *Parker* site are consistent with the sizes needed to secure the keel timbers. The shorter drift bolts from the site were likely used to attach the 30-cm square floor timbers to the keel, connect the 30-cm square paired frames together, or secure the 40-cm square keelson through the floors and into the keel (Hegarty 1964:30–31). Further evidence of the function of these bolts is found in a table of fastener dimensions published by David Leigh Stone (1993:36), which provides the diameters of different types of metal fasteners relevant to the tonnage of a ship. This table indicates that 2.8-cm bolts fit the diameter prescribed for keel, keelson, rider keelson, and deadwood bolts (all drifts) of a 350- to 450-ton ship. Those dimensions are consistent with the measurements recorded at the *Parker* site.

Three smaller fastener types were identified among the remains of *Parker*. A total of 34 fasteners used to attach the ship's pintles to the rudder were found intact within the four pintles (discussed in the next chapter). Known as rudder nails, these roughly 12.5-cm copper-alloy fasteners were also used for attaching the gudgeons to the rudder post by driving them into the timbers via premade holes (McCarthy 2005:174). A small number of spikes commonly used for attaching planking to the hull were also identified scattered around the site; these are approximately 15 cm in length, and their bent and twisted appearance indicates that they were in use when the ship wrecked. Copper tacks were the other type of fasteners found on the site. Measuring roughly 2.5 cm in length, at least 70 of these tacks were identified, and all were found still attached to sections of copper sheathing.

Evidence of sheet copper was recorded in Section B of the *Two Brothers* shipwreck site. Interestingly, of the six sheathing fragments identified, only one was recorded in situ. This fragment—approximately 7 cm in length—was situated on the seabed among a concentration of copper-alloy spikes. In fragile condition and with no indication of fastener holes noted, it appeared to be either part of a single piece that was folded onto itself during the wrecking event or portions of more than one sheet that could have been stacked together for storage. Outfitting records indicate that all early to mid-nineteenth-century whaleships carried extra sheets of copper for use in case of emergency (Kirby 1860; NBWM 1832; New Bedford Free Public Library Whaling Collection [NBFPL] 1841, 1845, 1851). The five other fragments of copper sheathing were extricated from a concretion that had formed around an iron cooking pot that was recovered for display (discussed in chapter 6). These small pieces were discovered during mechanical cleaning of the pot while it was undergoing conservation treatment and were stabilized as part of the process (Fox 2012:6). Although it is likely that this copper sheet was used or intended for treating the exterior of the ship's hull, it could also be associated with parts of the tryworks or possibly even the copper tank used for cooling whale oil.

Two sections of copper or yellow metal sheathing were also identified among a large concentration of artifacts resting in a pile near the northeastern edge of the *Parker* site. Associated with these fragments are numerous closely spaced sheathing tacks. The presence of in situ tacks may be an indication that the section of wood to which they were attached deteriorated in place. One piece, measuring roughly 49 cm in length and 5 cm in width, is likely a fragment of a plate, and its ragged edges suggest that it was sheared off at some point. The other is 90 cm in length and 8 cm in width and has sheathing tacks running along both long edges. The standard size for copper sheets of the mid-nineteenth century measured 35.5 cm in length by 122 cm in width (Hegarty 1964:49; Ronnberg 1985:183). It is generally accepted that these sheets were applied to a ship's hull as a full sheet with the overlapping edge running on top of the next sheet aft. Thus, it is likely that this section is what Hegarty (1964) refers to as a filling piece, which, along with wedge-shaped "goring pieces," were added where necessary to fill gaps (Hegarty 1964:50).

Rerigging

Once hull repairs and modifications were completed and the ship was again on an even keel, the next step was rigging. The task of rerigging involved an initial inspection of the masts, spars, and bowsprit and their associated ironwork to determine condition; components deemed unsatisfactory for further use were repaired or replaced. Though time consuming, this step was necessary to ensure that ships were well equipped to withstand whatever conditions they might encounter at sea.

To carry out the various tasks involved in rerigging, skilled craftworkers who specialized in particular trades were contracted. And though the wages paid for the different services varied greatly, each was performed with attention to detail (Hegarty 1964:129). Primary among these craftworkers were the ship smiths. These were essentially blacksmiths who specialized in nautical fabrication and were employed to repair or manufacture anew all "the bolts, straps, trusses, bands, hooks, pins and shackles upon which the life of the ship depended" (Hegarty 1964:129). Riggers worked closely with block makers who crafted the deadeyes used to stay the masts and blocks for the running rigging. They also coated the lines with tar to make them more resistant to the damaging effects of seawater before running them aloft and securing them (Verrill 1916:60).

Because the intense environmental conditions experienced at sea battered the sails and cordage, as a rule they were completely replaced and overhauled after each voyage (Baker 1974:3). As mentioned, the predominant sail configuration in the early nineteenth century was the full-rigged ship; thus, an enormous amount of canvas was required to manufacture sails. Since laying out and cutting each sail necessitated such a large floor space, the attics of buildings—referred to as sail lofts—were generally used for this purpose (Hegarty 1964:99). Once they were properly sized, sailmakers and their apprentices hand stitched iron attachment points such as clews, eyelets, and cringles to the new canvas where necessary. When ready they were "bent" or attached to the spars (Hegarty 1964:99; Paasch 1885:147). Extra canvas and ropes were also purchased to be used as replacements; these were prepared and then stowed in casks with their contents marked so that they could be accessed as needed (Baker 1974:3).

Archaeological Evidence for Rerigging

Archaeological material associated with the rigging of early nineteenth-century whaleships was documented at both the *Two Brothers* and *Parker* shipwreck sites. As with the artifacts related to ship architecture, none of the organic components survived in the waters of PMNM. Instead, the remains found at these sites consisted mainly of metal fittings from the ship's rigging. Among the types identified were mast and bowsprit fittings, ironwork associated with components of standing rigging, chain, portions of wire rope, and blocks used for running rigging.

A small number of iron hoops found among the main concentration of artifacts at the *Parker* site were interpreted as "mast bands." Full-rigged ships such as those used for pelagic whaling employed three wooden masts; each of those consisted of a lower mast of white pine (*Pinus strobus*) and two smaller spruce (*Picea* sp.) timbers known as topmasts, which were used to extend overall heights and add sail area (Hegarty 1964:94). Each mast and topmast included numerous iron collars called mast bands, which were heated and driven down on the mast to provide reinforcement (Campbell 1974:113). The number of bands depended on the size and composition of the mast, and some of them were fitted with eyes or swivels for attaching lines or tackles (Campbell 1974:113; Stone 1993:70). Although these artifacts are encased in marine growth and concretion, no eyes or other attachment points were noted on any of them. Based on their location and size, these are considered to have been associated with the foremast.

Other important rigging elements include "mast caps" or "bowsprit caps." These iron fittings were added at the junction of two mast or bowsprit timbers. To provide the required support for partnering the two timbers, they overlapped and were held in place by mast caps. These pieces were configured with a square hole and a circular hole on either side of a frame; since the head of the lower of the two mast or bowsprit timbers was intentionally squared, the square side of the mast cap was fitted over it; the bottom of the upper mast timber was rounded, and therefore the circular portion was fitted over it (Kipping 1859:26; Stone 1993:62). Earlier designs—referred to as "ironbound wooden caps"—employed solid blocks of hardwood with premade holes for the two masts and a thick iron frame fitted around the outer edge for reinforcement (Campbell 1974:115). Three rectangular iron bands with rounded corners were recorded at the *Parker* site and interpreted as being mast caps of this early design. The largest of these is mostly intact and likely joined the lower foremast to the top mast. The others are smaller, but similar in design, and likely connected either topgallant and/or royal masts or the two bowsprit timbers (Doane 1987:270–274).

Attached to the masts were numerous spruce timbers known as yards or yardarms, which hung crossways to carry the many different sized square sails that propelled the ship (Hegarty 1964:94; Stone 1993:64). The length of each of these spars varied with its position, and they were attached to the masts using iron fittings known as "parrels" or "trusses." These strong iron pivots were attached to the center of the yard and allowed it to move to trim the sails and be braced as required (Doane 1987:274; Kipping 1859:112; Stone 1993:65). The remains of four iron trusses were recorded at the *Two Brothers* site. Although heavily concreted, each of these consists of the thick central piece of the fitting, which is basically a U-shaped iron bar with a short arm extending down from the center (Figure 5.1). Although only one of these was identified in Section A, the three others were identified in proximity in Section B, suggesting that the mizzen mast—farthest aft from the bow—came to rest and deteriorated in that area.

Figure 5.1. One of three iron trusses documented at Section B of the *Two Brothers* shipwreck site. (Courtesy of Papahanaumokuakea Marine National Monument.)

As with masts, yards were reinforced along their length with iron bands, many of which were equipped with one or more eyes for attaching blocks or lines. Known as "cranse irons," these fittings were also used for creating attachment points on the bowsprit, which was a spar that projected forward of the ship from the bow and angled slightly upward (Stone 1993:60–65). Other fittings that were often added to the yards were "boom irons," which were metal rings fitted on the yards and used to connect extra sails in light winds. To connect the timbers, the lower hoop of a boom iron was slipped over the outer ends of the yardarm, and then a small boom was inserted through the upper hoop. From this extension, small sails called "stunsails" or "studding sails" were flown to add extra surface area (Canadian Parks Service 1992:59).

Several fittings associated with either the foreyards or bowsprit were identified at both of the shipwreck sites. Two cranse irons encased in dense marine encrustation were found at the *Two Brothers* site. These specimens included one with a single eye that was probably used on of the upper yards and a larger three-eyed band that may have attached the yard to the main foremast. Two cranse irons were also recorded at the *Parker* site; one with a single eye that was likely positioned near the center of a spar and one heavily concreted two-eyed band that attached toward the outer end of one of the upper spars. Two other metal yard or bowsprit fittings were identified among the remains of *Two Brothers*; however, their exact functions are undetermined. These unidentified artifacts are speculated to be boom irons; one of which resembles a quarter iron that would have attached "3/16 the length of the yard form the outer end" (Kipping 1859:28–29), while the other could be a yard arm iron that would have attached to the end of the spar and had another band projecting down form it.

A relatively large amount of chain was also recorded at the *Parker* shipwreck site, including numerous lengths comprised of two differently sized links. The presence of chain among the remains of a ship lost in 1842 is not uncommon since chain was regularly used for anchoring merchant vessels by the 1820s (Stone 1993:12). Not long after being introduced, mariners realized the versatility of chain onboard ships, and soon, it was used for many purposes other than as cable for anchors. Chains of smaller sizes were particularly favored for staying the bowsprit to the bow of the ship. Since the smaller of the two sizes of chain recorded at the *Parker* site is not considered to be thick enough to have operated as anchor cable and other fittings associated with the bowsprit rigging are present, it is possible that it was used for "bobstays," or the guy lines that connect to the stem to steady it (Stone 1993:61).

Perhaps the most prevalent of the rigging-related artifacts at both sites were deadeye strops and chain plates. Important components of a ship's standing rigging, these were used for attaching and tightening the shrouds, or ropes that supported the masts (Biddlecombe 1925:7; Davis 1918:80). Deadeyes were flat, circular blocks of hardwood such as elm (*Ulmus* sp.) or lignum vitae (*Lignum vitae*) that had a grooved outer edge and three to four holes piercing their faces. They were firmly bound within either an iron strop or the end of a rope, and they were used in pairs to secure the ends of shrouds and stays (Biddlecombe 1925:11; Doane 1987:262; Stone 1993:71–72). A short section of rope called a lanyard was threaded between the deadeyes and used to stiffen the rigging when tightened (Stone 1993:71). Since whaleships often encountered treacherous conditions while at sea, it was vital that their masts were kept steady. Deadeyes proved to be the most effective method for doing so until the 1860s, when turnbuckles began to replace them. Several different sizes of deadeyes were used at various points in both the upper and lower rigging, with the largest of them added to chain plates to support the masts (Stone 1993:72).

Chain plates, also referred to as simply "chains," were iron rods that were bent in such a way as to create a closed loop that incorporated a small diameter hole at the bottom end and a larger diameter hole at the upper end (Hegarty 1964:88). They were bolted to the topside portion of the outer hull planking through the small hole on the lower ends and reinforced using smaller iron rods known as preventers (Desmond 1919:208). Chain plates were usually bent out from the side of the ship and run through a thick, horizontal oak plank called a channel, which kept them clear of the bulwarks (Davis 1918:80; Steffy 1994:269; Stone 1993:72). A whaleship generally had six or eight chain plates located abreast of each mast on each side of the vessel (Figure 5.2). The shrouds attached to the tops of the masts included a deadeye at their lower ends that attached to the tops of the chain plates, which were also equipped with deadeyes, to work as vertical pairs to stiffen the mast (Stone 1993:71). Ratlines, or small shots of evenly spaced rope, were placed horizontally above the forward pair of chain plates of each mast, and whaleboat crews used them to easily climb up and over the bulwarks (Biddlecombe 1925:23; Higgins 1927:15).

Eleven possible deadeye strops or partial chain plates, as well as two complete chain plates were documented at the wreck of *Two Brothers*. All of these were identified on the reef flat within a portion of Section A and were either concreted together in a small pile or found resting among the broken coral on the seabed (Figure 5.3). Since many of these artifacts are damaged or are heavily concreted into the reef structure, determining whether they are partial chain plates was not possible. Several smaller deadeye strops—some of which are connected to short loops of iron rod—were also identified. Based on their sizes, these are interpreted as having been used in the upper rigging for steadying topmasts. Although concretion and marine growth distort their actual dimensions, the inner diameters of the holes intended for deadeyes on these artifacts ranged from 10 to 19 cm, which is consistent with the deadeye measurements offered by historian Reginald Hegarty in *Birth of a Whaleship* (1964:88–89). The spatial location of all of the rigging components within the *Two Brothers* site supports the interpretation that the ship foundered on the reef top after its stern was smashed open and that the masts and rigging came to rest on the backside of the reef.

Figure 5.2. Chain plates attached to the midship portion of the hull of the whaleship *Charles W. Morgan* and used to stay the mainmast. (Jason T. Raupp.)

Figure 5.3. Pair of deadeye strops documented at Section B of the *Two Brothers* shipwreck site. (Courtesy of Papahanaumokuakea Marine National Monument.)

A smaller number of deadeye strops and chain plates were identified among the main artifact concentration at the *Parker* shipwreck site. As with the standing rigging components recorded at the *Two Brothers* site, the thick concretion and coating of marine growth distorts the actual dimensions of these artifacts and makes determining their actual functions difficult. Only three of them were determined to be mostly intact chain plates, although portions of other broken deadeye strops are also present. Identifying features for these chain plates include the large hole intended for a deadeye to be inserted and the section of iron rod that extends below. These examples range from 1.6 to 2 m in length. Based on their location and other remains found in this area, these were probably associated with the foremast.

Other remains of standing rigging identified at the *Parker* shipwreck site include portions of wire rope. Sections of this fragile material were found scattered among the main concentration of artifacts with lengths ranging from 9 cm to approximately 3 m. All portions recorded are heavily encased in calcareous marine encrustations; two small samples recovered in 2005 revealed that the wire inside the encrustation is extremely fragile, with one of them having mostly deteriorated to a black sludge (Fox 2006). Based on its location and proximity to other standing rigging pieces, such as chain plates and deadeye strops, this wire rope could have been used for shrouds to secure the foremast or for forward stays. The presence of wire rope onboard an American vessel of this date is peculiar since it was only developed in England in the early 1830s. The durability, lighter weight, smaller size, and economy of this material, however, led to its rapid adoption onboard both British naval and merchant vessels (Martin 2014:153; Stone 1993:69–70). Although it did not begin to supplant manila ropes on American ships until the 1860s, patented wire rigging is known to have been in use on some American vessels by 1840 (Van Tilburg 2003).

Three rigging blocks were also documented at the wreck of the whaleship *Parker*. Blocks consist of a frame within which one or more grooved pulleys are mounted on a pin; ropes are passed through the frame and over the pulleys to increase lifting or holding power (Biddlecombe 1925:3; Doane 1987:258). As with all sailing vessels of the early nineteenth century, numerous blocks of varying size and configuration were used throughout the rigging for lifting and lowering everything from sails to cargo and for tightening guy lines (Doane 1987:258; Stone 1993:71). A review of various preprinted outfitting books dating from the 1830s to the 1860s reveals several different types of blocks and spare parts listed among the ship's complement; these include burthen blocks, purchase blocks, guy blocks, and cat blocks (NBWM 1832). Although heavy marine encrustation obscures much of them, each was recorded to note size, style, and components. Ranging in overall size from 23 cm to 57 cm, all three artifacts appear to represent iron-framed single blocks with at least a partial sheave intact and one with a hook connected to one end. Since each of these was recorded among the main concentration of artifacts, they were probably associated with *Parker*'s foremast.

Other rigging elements identified at both shipwreck sites were thimbles. These were iron rings with grooved outer faces for seating a line that were inserted into eyes made in ropes; once in place the line was then spliced back into itself around the ring or "seized" by tightly binding the end back onto itself using smaller line or "spunyarn" (Biddlecombe 1925:26–34; Davis 1918:120). Thimbles of varying sizes could be found connected to the rigging in many places on any working whaleship of the period. Large quantities of thimbles were also shipped onboard as spares. Outfitting records for the New Bedford whaleship *William Rotch*, for example, indicate that they were purchased by the pound rather than actual numbers because they were so numerous (NBWM 1819). Being such ubiquitous objects on sailing ships, it was unsurprising to find numerous thimbles scattered among the remains of both *Two Brothers* and *Parker*.

Topside Refit and Painting

Carpenters found no shortage of work to keep them occupied during the refit; below deck they renovated the cabin and forecastle and inspected all interior spaces for rotting timbers (Littlefield 1906:10). On the whaleship's topside, they checked the condition of deck furniture, such as hatch and companionway coamings, the skylight, fife rails, and stanchions, and repaired or replaced them when necessary (Stackpole 1967:31). By the 1860s, deckhouses became prominent features at the after end of the whaleship, and carpenters would also attend to them; however, for the period under consideration, no such structures were incorporated since a flush deck was considered more convenient by most captains (Doane 1987:61; Leavitt 1973:14). Painters were kept busy coating nearly every surface of the vessel, inside and out, with fresh paint. The hulls of most ships were painted black from the upper edge of their copper sheathing to the tops of their gunnels, and a white band was generally added around the ship above the water line (Decker 1974:29). Onto this white band false gunports were often painted in black to trick would-be attackers into thinking that the ship was armed (Dickerman 1949:5; Hegarty 1960:37; Littlefield 1906:10; MacGregor 1989:87).

Perhaps the most important job when undertaking the topside refit was the thorough inspection and report of the ship's windlass. Although all merchant ships of the period were equipped with a windlass to assist in hoisting anchors (Desmond 1919:211), by their nature whaleships had little use for an anchor since they spent several months in water that was far too deep to consider using one. Regardless, the average whaleship left port with several anchors onboard; among them were the two large bower anchors rigged on the bow of the ship and raised using the windlass, as well as the smaller and lighter ones called kedge anchors that were kept in storage and used to move the ship from one berth to the other or clear the ship if it ran aground (Jobling 1993:139).

Windlasses used on whaleships in the early nineteenth century were little more than two wooden uprights, a barrel, and a handspike. As such, using them required considerable time and exertion (Mawer 1999:254; Stone 1993:42). By the 1820s, however, the introduction of the geared windlass greatly improved its efficiency (Davis et al. 1997:272). Known as pump-brake windlasses, these new models required four or five crew members on each side to operate by alternately pumping on a double-handled rocker lever that ratcheted the gear on the center of the barrel and increased tension (Grant 1932:32). A metal piece known as a pawl dropped into the teeth of the gear as it passed and kept the barrel from unwinding upon the release of pressure (Stone 1993:42). The mechanical design of the ratchets and more effective positioning of the brakes resulted in increased power (Chapelle 1973:677) and significantly eased the task of lifting heavy objects.

Archaeological Evidence for Topside Refit and Painting

The remains of a pump-brake style windlass were identified at the *Parker* site. Although the structural components of this machine were constructed of wood that has long since broken down and deteriorated, several of the metal fittings associated with its revolving part, known as the barrel, are present at the site (Hegarty 1964:55). These include the roughly 8-cm square iron shaft on which the wooden barrel was built; the large, circular iron gears known as purchase rims; and the multiple iron bands that were used to help bind the wooden barrel (Baker 1984:56; Chapelle 1973:677). Despite being heavily coated in marine encrustation and concretion, all these components appear to be relatively intact and in stable condition. The purchase rims and some of the bands still stand relatively upright in their original positions on the barrel, and the metal shaft rests within them. Although the exact size of *Parker*'s windlass is unknown, the central part of the barrels equipped on comparable-sized American whaleships of the period measured 3–3.4 m (Hegarty 1964:55). Based on the context of these upright purchase rims, the windlass remains found at the site indicate an overall barrel length consistent with those measurements.

Closely associated with the windlass were the ship's anchors and their hardware. Three anchors were recorded among the remains of both the *Two Brothers* and *Parker* shipwrecks. These artifacts are all considered to be very well-preserved examples of their types, and the styles of those at each site are indicative of changes in anchor technologies that occurred between the 1820s and 1840s. Also documented at each site were some components associated with either anchors or the anchoring process; these include anchor rings, shackles, and hawsepipes.

Two of the three anchors identified at the *Two Brothers* site fit the description of "old pattern long shank anchors" due to their pointed crowns and the angles of their arms (Rubin 1971:231). Naval and merchant ships of the eighteenth and early nineteenth centuries commonly used this anchor design (Nash 2009:136). Although there was some variation in form between the different countries that produced them, for the most part these were minor (Rubin 1971:231), and the basic style is commonly used as a diagnostic feature. The two large anchors recorded at the *Two Brothers* shipwreck have approximately the same dimensions—3.65 m in length and 1.9 m between the bills—and are interpreted as bower anchors, or the ship's principal anchors that were carried on the bow and generally rigged and stowed for immediate use (Jobling 1993:139; Steffy 1994:266). Although both anchors are in Section A of the site, they are separated by over 170 m. This distance can be taken as an indication of the violence with which the ship broke apart and scattered across the reef. A diagnostic feature of these anchors is the large iron "ring" that was attached to the eye of the shank and used for fastening the cable (Jobling 1993:138). The presence of rings indicates that the anchors most likely predate the introduction of the shackle in the 1820s.

Although neither stock of the bower anchors remain, the presence of a small notch on two sides of their shanks indicates that the stocks were wooden. Known as the stock key, this feature locked the stock in place and prevented it from pivoting around the shank. Stocks were crosspieces mounted near the top of an anchor's shank and placed perpendicular to the arms to allow them to "cant" or turn and dig into the sediment (Jobling 1993:139; Steffy 1994:267). Later anchors used metal stocks, but earlier examples were constructed of two pieces of wood that were joined around the shank and held together using fasteners and bands called hoops (Jobling 1993:139). Though the name refers to a round shape, these iron straps were often square. Generally, two hoops were used on each side with the outer being smaller in diameter since the wooden stock tapered out from its center (Jobling 1993:137–139). A set of anchor hoops was recorded at the *Two Brothers* site, though they were probably not associated with the bower anchors. Instead, their location in Section B indicates that they were likely carried as spares and were stored in the stern at the time of loss.

The other anchor recorded at the *Two Brothers* site is likely a kedge anchor.

This type of anchor was used for centuries to perform duties such as moving a vessel, temporarily holding one in a waterway, or clearing one that had run aground (Jobling 1993:139; Steffy 1994:267). Similar in shape to bowers, though much smaller in size and lighter in weight, kedge anchors were generally kept stowed away and only deployed when needed (Desmond 1919:157). The possible kedge anchor found at *Two Brothers* measures 1.85 m in length by 1.05 m between the bills and has a metal stock. Although it is commonly thought that metal stocked anchors were not introduced until the early 1800s, Steel's *Elements and Practice of Rigging, Seamanship, and Naval Tactics* indicates that iron-stocked kedges were in use by at least 1794 (Jobling 1993:110–113; Steel 1794:80–81). The kedge recorded at *Two Brothers* has an eye at the top of the shank, but no ring is present; however, an incomplete anchor ring 30 cm in diameter and recorded nearby could be associated with it.

Two bower anchors identified among the main concentration of artifacts on the *Parker* shipwreck site are associated with the ship's bow. Based on their curved rather than pointed crowns, these were initially identified as being "Admiralty style" anchors (Van Tilburg 2005). As this anchor type was only introduced to the British Navy in the same year that *Parker* was outfitted (1838), it is more likely that these are of an earlier type on which the Admiralty pattern was based, such as those designed by Richard Pering around 1830 (Rubin 1971:231). Further evidence for this determination is provided by the dimensions of both; respectively these anchors measure 3.1 m in length by 2.6 m between the bills and 2.85 m in length by 2.4 m between the bills (Figure 5.4). According to George Costell's *A Treatise on Ship's Anchors*, Pering's designs altered the proportions of the anchors such that they were known as Pering's Improved Anchors with "Short Shanks and Long Arms" (Costell 1856:12). As with the bower anchors recorded at *Two Brothers*, these employed wooden stocks that were held together using fasteners and hoops. Two hoops were identified at the *Parker* site. Based on their proximity to one another and the fact that one is larger, these appear to comprise one side of the anchor stock for the westernmost of the two bower anchors.

One other anchor was identified among the remains of the whaleship *Parker*. This artifact was identified partially buried in the main section of the site and measures approximately 84 cm in length by 54 cm between the bills. No stock is evident, and the crown is elliptical; small palms are present at the end of each arm. Although lighter, four-armed grapnel anchors are generally listed as the type used in whaleboats; some ambiguity exists in outfitting records. For instance, none of the outfitting books reviewed lists grapnels as being included with boat equipment; however, most include at least one "boat anchor," which was shipped onboard by the 1830s (Kirby 1860; NBFPL 1840, 1841; NBWM 1832).

Figure 5.4. One of two bower anchors documented at the *Parker* shipwreck site. (Courtesy of Papahanaumokuakea Marine National Monument.)

Rather than using rings for attaching to the cable, iron shackles were documented at the upper end of each anchor at the *Parker* site. By the 1820s, chain began to replace hemp rope as the preferred cable on merchant ships, and shackles were used to connect chains to the anchor (Stone 1993:12). Shackles are composed of a U-shaped link with holes at its ends and a pin, which was inserted in the holes to create a closed loop; these were connected through the eye at the top of the anchor's shank (Stone 1993:7). Though numerous sections of chain that were likely used for anchoring the ship were identified at the *Parker* site, none of it appears to have been attached to the anchors when the vessel wrecked. Since the ship was accidentally lost on the reef while transiting through the region, this lack of attachment to chain could indicate that the anchors were disconnected and stowed while the ship was at sea.

One other artifact type associated with the anchoring process found on both the *Two Brothers* and *Parker* sites were lead hawseholes. Hawseholes were tubes that lined the holes made in the ship's bulwarks and planking to allow the anchor cables to pass and protected the timbers surrounding them against damage from chafing (Steffy 1994:272; Stone 1993:38). According to Stone (1993:37), the presence of these artifacts is unmistakable proof of the bow portion of the wreck. Two hawsepipes were recorded among the remains of *Two Brothers*, and both were located in pockets in the reef between the site's two bower anchors in Section A. The one identified at the *Parker* site was lying directly in front of one of the two bower anchors and next to the remains of the ship's windlass. Similar artifacts have been identified on the wreck of the British whaleship *Lively*, lost in 1811 off western Australia (Henderson 2007:97), as well as the *Mica* wreck, an early nineteenth-century merchant schooner lost in deep water in the Gulf of Mexico (Jones 2004).

Industrial Elements Added

Also added during outfitting were some of the vessel's industrial elements. These features included the davits on which multiple whaleboats hung, hull and rigging modifications required to accommodate carcass flensing, and the large brick tryworks used for boiling blubber. Once added, these features altered the profile of the ship and made it unmistakable to those encountering it (Brown 1887:234). Unique to whaleships, these features evolved with the sperm whale fishery; as seen in the discussion of the industry's historical development, the whaling vessel underwent a marked evolution with the expansion into deep-sea hunting in the early decades of the eighteenth century (Tower 1907:86). Each phase in that development brought the need for larger and more complex vessels that could withstand the increasing pressures associated with greater distances traveled, which in turn helped to formalize the system for catching and processing whales and standardize the equipment employed. While these components are among the most noticeable in the archaeological record, those remains found on the *Two Brothers* and *Parker* sites ware discussed in detail in the following chapter.

The first industrial element to be added was related directly to the hunting and capture of whales. Because early nineteenth-century ships were far too slow to actively chase whales, ships generally carried between three and five small, fast whaleboats for that purpose. The practice of employing a "mother ship" and chase boats can be traced back to the earliest sperm whaling cruises of the seventeenth century by single-masted colonial sloops (Fonda 1969:24–25; Stackpole 1953:31). Though faster than nineteenth-century whaleships, the seventeenth-century Nantucket sloop also proved impractical for taking whales due to its size. To overcome this issue, small boats used in the shore fishery were hoisted onboard and carried on deck on wooden arms referred to as "tail feathers" that projected from the stern or were simply towed if conditions allowed (Fonda 1969:24–25; Hall 1884:23; Kugler 1980:7). Boats kept onboard were deployed using curved or straight-armed wooden cranes called davits that suspended the boats over the sides of the mother ship without interference (Doane 1987:262). The inclusion of whaleboats onboard a larger vessel was the first significant step in the establishment of a standard form for the pelagic whaleship, and from that point, one could always be recognized by the numerous boats hanging on davits (Hegarty 1964:81).

By the early nineteenth century, the system for launching boats to chase whales from the ship had been perfected, and davit design was standard. Built of a hardwood such as white oak (*Quercus alba*) or elm, the davit timbers stood at least 3 m high and were positioned at specific stations along the sides of the ship. Each of these had a particular name such as aft, waist, or bow boat that corresponded to their position on either the starboard or larboard (port) rail (Brown 1887:243; MacGregor 1989:87). Each boat required two davits, each of which was rigged with blocks for raising and lowering them (Figure 5.5). Each station also utilized other timbers to form a cradle for supporting the boats when raised. Those timbers included two uprights called bearers, which were stanchions placed between the davits and onto which the whaleboat's gunwale rested (Ansel 1983:142); triangular wooden brackets called cranes, which were attached to the bearers using pintles and eyebolts and swung out into a horizontal position so the whaleboat could rest on its keel when raised (Brown 1887:243; Leavitt 1973:122–126); and vertical slide boards, which were light, springy boards attached to the ship's side from the rail to the waterline to prevent whaleboats from being damaged from rubbing (Brown 1887:243–244; Hegarty 1964:80–81; Leavitt 1973:122–126). Once raised into the cradle, the whaleboats were secured using an iron rod that was attached to an eye on the side of the cradle arm and was hooked to an eye on the boat, as well as a rope called a gripe, which was passed from the ship, underneath and then over the boat, and back to the ship where it was made fast (Doane 1987:65). Though iron davits were increasingly used onboard whaleships by the mid-1840s, many whaleships continued to use the wooden form throughout the nineteenth century (Doane 1987:262).

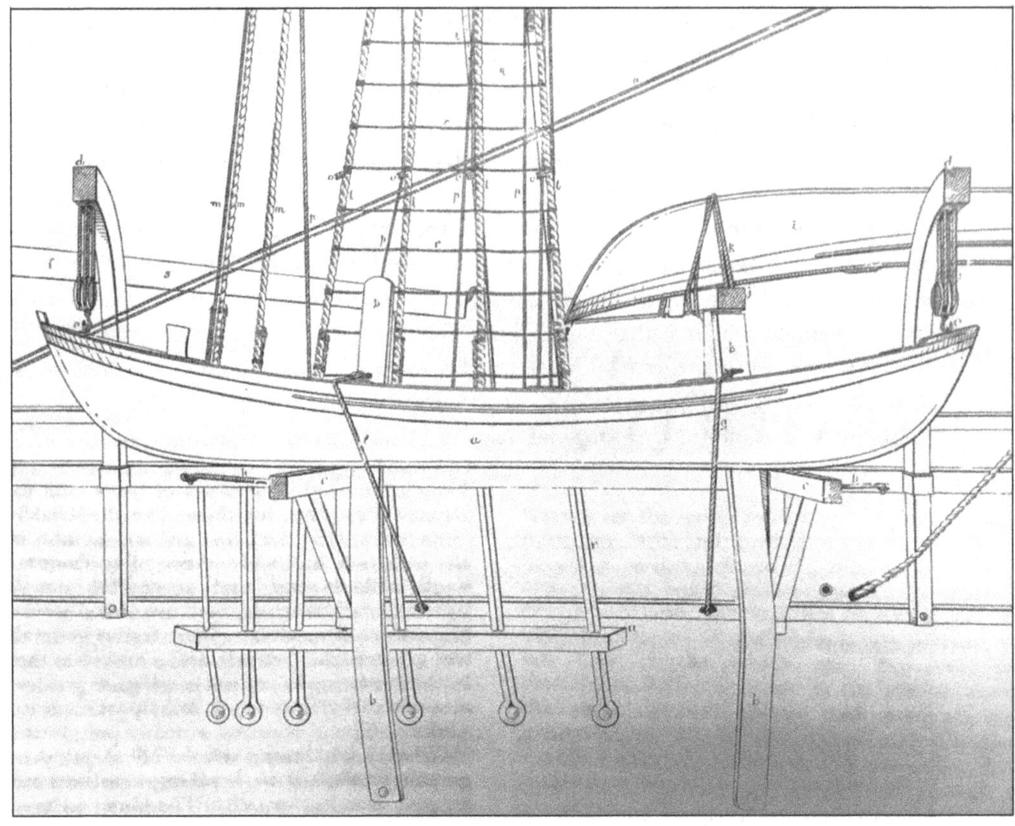

Figure 5.5. Whaleboat (*a*) resting on cranes (*c*). Note the slide boards (*b*) attached to the stanchions located just to the inside of the cranes, as well as the blocks and tackles from the davit heads (*d*) for raising and lowering. (From Goode 1887.)

Along with the installation of the davits came the need to procure whaleboats. Since the ship was too large and slow to actively chase whales, whaleboats were among the most important factors in the capture of whales (Brown 1887:240). By the 1820s, a well-equipped whaleship usually carried four or five whaleboats on davits and two or three spare boats lashed to a simple wooden framework called a "skids," which was erected roughly 2 m above the deck between the main and mizzenmasts (Brown 1887:245; Littlefield 1906:8; Nordhoff 1874:44). Because of the long duration of the cruises and the many perils involved in hunting whales, whaleboats were often damaged or destroyed; thus, most ships also carried "knocked down" boats, the components of which were prefabricated to be assembled when needed. With the increase in the number ships engaging in sperm whaling after the War of 1812 came a great demand for quality whaleboats. Thus, it was important to engage local shipwrights early in the outfitting process so that the desired number of boats would be ready by the intended date of departure.

Between 1750 and 1800, the boats used onboard American whaleships evolved to an apex form that changed little over the following century (Kugler 1971:21). Designed to be as versatile as possible, these were relatively lightweight, narrow, smooth-bottomed craft that offered both great velocity and safety in rough seas (Hall 1884:23; Olmsted 1841:19). Sharp at both ends so that they could be rowed from either direction, they measured approximately 8.5 m in length, 1.8 m in beam at amidships, and 0.6 m in depth (Brown 1883:6; Davis 1874:158; Morrison 1921:318; Olmsted 1841:19; Spence 1980:101; Starbuck 1878:123). Oak, spruce, and pine were used in building the boats, and cedar was preferred for planking due to its qualities of being lightweight but tough when wet. The planks were attached in a lapstrake construction technique (i.e., overlapping hull planks), and all timbers were connected using copper fasteners. By 1820 it was common to paint these boats white with their sheer strakes green, black, or blue (Brown 1887:241; Davis 1874:157; Hall 1884:23; Morrison 1921:318; Starbuck 1878:123–124). The few modifications made to these boats after the middle of the nineteenth century included the addition of a centerboard and a change in the planking assembly to employ a combination of clinker (lapstrake) and carvel (edge-to-edge) methods (Ansel 1983; Kugler 1971:21).

Whaleboats of the early nineteenth century were propelled either by rowing or sailing. Each boat was equipped with five thwart timbers on which the boat header, boatsteerer, and rowers sat; each of them had specific names depending on their position and had a corresponding cleat for peaking oars and either tholepins or an oarlock set into the top of the gunnels. These fixtures helped to control the oars and were coated with soft rope to muffle the sound to keep from alarming whales of their presence (Davis 1874:157; Doane 1987:65; Edwards and Rattray 1932:55–56; Higgins 1927:9; Starbuck 1878:123). The lengths of the oars varied depending on the position, and all rowers were also furnished with a small paddle kept under the thwart for use when approaching whales (Doane 1987:65; Olmsted 1841:19). The sail configuration was simple; it employed a single mast that could be raised and lowered as needed and generally a single fore-and-aft rigged sail pattern (Kugler 1971:21).

Aside from the thwarts, the features of the interior of whaleboats were mainly at the ends. In the bow was a box for coiling short ropes or warps, a plank called a "clumsy cleat" that had a section cut out for the harpooner to rest his thigh when approaching a whale, and an aperture on top of the stem called a chock, through which the whale line passed (Doane 1987:65). The aft section included a heavy post known as a loggerhead, which was used for managing the whale line; a projecting sternpost, to which a strap was attached for keeping the 6.1 m long steering oar in place; and a space large enough for the boat tub, into which was coiled 180 to 270 m of whale line for attaching to harpoons (Ansel 1983:14; Davis 1874:157; Doane 1987:67; Olmsted 1841:19; Starbuck 1878:124). The inside of the boat was covered with ceiling planking, and at each end, a small, slightly raised platform was added to the floor; the one forward was called the bowsheets and the one aft the sternsheets (Doane 1987:67).

Other modifications and additions to the ship's hull during outfitting were associated with stripping the whale's carcass; this process was called "cutting in" by American whalers, "flensing" by British, and "flinching" by old Nantucketers (Brown 1887:277). From the advent of sperm whaling, American whalers brought the carcass alongside the ship for flensing (Fonda 1969:10), and over time, a standard practice developed that required specific modifications to the ship. Weather permitting, cutting in was always done at the starboard gangway on the windward side where a 3 m long section of bulwark could be removed (Leavitt 1973:23). This detachable section provided deck-level access to the carcass, and from it, a small platform known as a cutting stage was rigged so that flensing crews had access on both sides.

In the early nineteenth century, this projection was simple and could be rigged either as a one- or two-section configuration. The single section form employed one plank roughly the length of the removed bulwark section, which sat at a right angle to the deck and was supported by tackles and lashings from the rigging and rails (Leavitt 1973:123–124). The double section form used a smaller platform of similar design rigged on either side of the gangway opening (Ashley 1926:97; Olmsted 1841:62). On the outside of the hull just aft of the gangway, small cleats were attached one above the other to form steps so boat crews had easy access to the ship when the gangway was open (Higgins 1927:15).

Though a detailed discussion of the cutting in process is provided in the following chapter, it is necessary to explain some aspects of it here in order to understand other associated hull modifications. Once a whale carcass was near the ship, it was positioned so that the head pointed to the stern and the tail toward the bow (Scammon 1874:232). To hold the whale in place, it was fastened to the vessel using a chain that passed through small holes cut low through the bulwark and plank sheer forward of the foremast; the chain was then fastened to a heavy upright timber located abreast of the foremast on the main deck called a "fluke bitt" (Doane 1987:63; Leavitt 1973:124; Stackpole 1967:31). The fluke bitt was a square hardwood post that projected above the deck and had an iron pin running athwartship through its upper end (Higgins 1927:28). Either a section of heavy manila rope or a specially designed 2-cm iron chain fitted with a large iron ring at one end—called a fluke chain—was used for this purpose (Doane 1987:63). To prevent the rope or chain from damaging the wood of the starboard bulwarks, the holes were fitted with 10-cm-diameter, circular iron pieces called fluke pipes (Leavitt 1973:124), which had rounded lips on the outside edges and were reinforced on the inside of the bulwarks by a thick piece of oak (Higgins 1927:15).

As noted, an important component of the ship's industrial equipment was the windlass. By their nature, whaleships had little use for an anchor; they spent several months in water that was far too deep to consider using one. Since the primary purpose of the windlass was raising and lowering anchors, it sat idle except on the occasions that the ship neared land. Given its large size and immense power, however, whalers developed a secondary use for it—lifting the heavy blubber pieces removed from carcasses onto the deck. It was this function that made the windlass an essential part of the whaling equipment.

Whalers used the mechanical capability of the windlass to their advantage by reconfiguring some of the rigging to enable the windlass to assist in getting blubber onboard and lowering casks into the hold. Rigging modifications were simple; a chain or rope pendant often called a necklace was attached around the main masthead, and rigging blocks known as "guy tackles" were attached to the fore masthead (Leavitt 1973:124; Scammon 1874:232). These additions were mostly unnoticeable and went unused until a whale was brought alongside the ship for cutting in; at that point a number of oversized blocks known as "cutting tackles" were attached to the necklace on the mainmast, and heavy manila lines called "falls" were run through them and down to the windlass for lifting blubber pieces (Leavitt 1973:124; Scammon 1874:232; Verrill 1916:41). Since windlasses played such an important role in the industrial operations onboard a whaleship, great care was taken to maintain them, and thorough inspection of the multiple wooden and iron components was included in a vessel's regular maintenance schedule while at sea.

The other main component of the industrial equipment added to the whaleship were the tryworks. This large brick structure was central to pelagic whaling operations and was used for boiling blubber into oil. Hailed as the most revolutionary innovation in the development of offshore whaling (Davis et al. 1997:36), the incorporation of the tryworks onboard American vessels transformed all aspects of the business. Prior to their installation, the vessels employed were "mother ships" of limited capacity that towed or carried smaller vessels to sea for a short time and returned with blubber in barrels for onshore oil extraction. With the introduction of shipboard tryworks, however, whaleships became full-fledged offshore production platforms where all processes necessary for the industry's primary resource extraction were conducted. Though Basque whalers first experimented with shipboard tryworks in the late sixteenth century, the climate of the Arctic grounds on which they hunted and the strategy they employed meant there was little need for them (Barthelmess 2009:662; Francis 1991:48; Morton 1982:30). It was only in the mid-eighteenth century, when American whalers moved into warmer waters to the south, that this technology was embraced fully.

So basic was the process of trying out, that the structure of tryworks changed little over the life of the fishery. The tryworks were essentially a brick fireplace into which large iron cauldrons called try-pots were situated (Hegarty 1960:75). Fires were built in furnaces beneath the pots to heat blubber until the oil separated from the fibers. When it reached a proper quality, the oil was then skimmed from the top (Figure 5.6) and put into a cooling tank (Francis 1991:48). The structure of the tryworks had multiple components, and, unless being installed on a newly built ship, building them first required that the deck be scrubbed and checked for rotten planks. Preparing the deck was made easier thanks to the common practice aboard nineteenth-century whaleships of "heaving the tryworks" at the end of a voyage (Stackpole 1967:60). More commonly referred to as "knocking down," this time-honored tradition involved breaking up the structure and casting the bricks and mortar overboard but retaining the pots and iron fittings (Ashley 1926:96; Chatterton 1925:171; Dakin 1934:4; Davis et al. 1997:247; Ellis 1991:143; Leavitt 1973:10; Palmer 1959:145; Van Tilburg 2002b:255). Doing this allowed the deck beneath them to breathe and dry out while a whaleship neared its home port.

The deck space needed for the tryworks was forward of the main hatch, and the size varied slightly—on average it was roughly 3 m in length by 2.75 m in width (Hegarty 1964:61; Littlefield 1906:10). Once the deck was prepared, the brick structure that encased the try-pots was constructed. Since the intense heat generated by the tryworks fire presented a major hazard on a wooden ship, the first step required the building of an enclosure that whalers called a duck pen or goose pen (Church 1938; Hegarty 1964:61; Weslowski 1978:3). This was made either by laying several courses of bricks or 7.6 cm thick planks around the perimeter of a base of 5-cm thick planks and then adding a layer of sand and bricks. When the tryworks were in operation, water was continuously circulated around this enclosure to keep the deck cool and wet (Ambrose 2000:90; Hegarty 1964:61). On this layer, brick supports for the pots and the walls were erected; the upper 30 cm of the walls was corbeled to fit the pots, and on early whaleships, the entire surface was covered in a layer of cement mortar that slanted toward the pots (Hegarty 1964:61). The terms "camboose" (Ambrose 2000:90; Macy 1835:228), "caboose" (Hazen 1854:84), and "caraboose" (Hawes 1924:172) have each been cited in whaling literature when referring to the entire brick structure; however, other authoritative sources (Hegarty 1960:32, 1964:61) use the term "camboose" to refer only to the water-filled enclosure beneath the fire. The word *camboose* is

Figure 5.6. Illustration of whalers loading blubber pieces into a three pot tryworks. (From Davis 1874.)

derived from the Dutch word *kombuis* (galley), and it also referred to a whaleship's galley, which was located near the stern (Baker 1973:212).

Though in later years the sides of the brick structure were sheathed with steel plates for reinforcement, in the early nineteenth century, wooden sheathing was used for the same purpose (Hegarty 1964:64; Weslowski 1978:3). The sheathing helped to strengthen the tryworks, but the massive structure was held in place by two heavy iron knees (known as tryworks knees) bolted to both the brick and the deck near the corners on each side (Hegarty 1964:62; Higgins 1927:25). On the forward edge of the tryworks were two 35-cm by 60-cm spaces that provided access to the fireboxes; these could be closed by sliding iron fire doors that hung from a heavy iron rod (Hawes 1924:172; Hegarty 1964:62). Above the duck pen on the inside of the structure were two square iron bars that supported multiple, smaller grate bars; these were used to make a base for the fire (Hegarty 1964:62). In the back part of the interior of the structure, the brickwork tapered to form flues that allowed the smoke to escape (Weslowski 1978:4).

Inside the tryworks were positioned two or three large iron cauldrons called try-pots. These pots were purpose-built, and those used in the early to mid-nineteenth century ranged in capacity from 140 to 200 gallons (530 to

760 liters) (Hawes 1924:172; Kirby 1860:5; Macy 1835:228; Weiss 1974:87). The try-pots had three pegged feet on the bottom and two lifting points positioned directly across from one another on the rim. Many of the pots used in the early nineteenth century had a flattened face on one side so that two pots could abut, ensuring optimum use of the furnace's heat (Pearson 1983:48). Since the try-pots were so heavily cast, it was seldom that they became defective or cracked; regardless, it was standard practice to carry a spare pot lashed to the starboard fore-corner of the tryworks (Hegarty 1964:63).

Another large pot carried onboard a whaleship was the oil cooler. This large copper or galvanized iron tank, lashed to the starboard side of the tryworks, was used for cooling the oil before it was run into the storage casks (Littlefield 1906:10). The oblong-shaped cooler was roughly 60 cm in width and 1.2 m in length, open at the top, and was approximately 6 cm lower than the top of the tryworks (Hegarty 1964:63).

Finally, there was a workbench—often called the cooper's or carpenter's bench—constructed on the after side of the tryworks (Doane 1987:62; Higgins 1927:28; Stackpole 1967:31). Though not an actual piece of the tryworks or directly associated with processing blubber, this bench was nonetheless important for cask construction and vessel maintenance activities and was added just after the tryworks were completed. The width of this six-legged bench was slightly less than that of the tryworks, its bench top was 60 cm in width, and it stood approximately 106 cm tall (Higgins 1927:28). As deck space was always at a minimum due to the massive amount of gear carried onboard whaleships, often the bottom of the bench was used as a coop by adding light strips nailed vertically to the three exposed sides (Higgins 1927:28; Whipple 1979:79). Livestock kept in coops onboard whaleships included pigs, fowl, and goats and were luxuries used for the officers' meals (Calkin 1953:31; Littlefield 1906:12).

Outfitting: Supplying the Ship

Once all inspections, additions, and modifications were complete, the ship was physically ready for the sea. Thus began the second phase of outfitting, which involved stocking the ship with all the supplies and stores needed for the upcoming cruise. Even though it was standard procedure to reprovision at friendly islands when supplies of fresh vegetables, fruit, and water ran low, agents did their best to outfit the ship to last the entire voyage (Allen 1973:160; Finney 2010:78). Outfitting was a huge task since it required an incredible range of materials including industrial equipment, resources for making repairs to the ship and whaleboats, and casks necessary for oil storage, as well as provisions, medicine, galley wares, and other items for use by the officers and crew (Finney 2010:233). In his detailed treatise on the American whale fishery, Captain Charles Scammon (1874:219) stated that "there are over a thousand different articles required to complete the outfit of a first-class whaleship, many of them of trifling value to be sure, yet all important to the success of the voyage."

As was the custom of every aspect of accounting in the whaling business during the golden age, the agent kept detailed records of all goods brought onboard. Once at sea, the master was responsible for managing the supplies as efficiently as possible and keeping track of materials used. When a voyage concluded, an inventory of all remaining equipment and provisions was completed; since much of the gear could be reused for subsequent voyages, this information allowed the agent to better determine actual profits by deducting those expenses. To assist in this operation, it was standard that an outfitting record listing all items shipped aboard the vessel was completed at the beginning of each voyage (Davis et al. 1987:35). Such meticulous recordkeeping can be traced back to the early pelagic whaling period, as is evidenced by documents on file in the Martha's Vineyard Museum archives. One such document is a handwritten list titled "Outfit for a Whaling Vessel in 1765" and was kept by the captain of the schooner *Lydia*, which sailed from Nantucket on a voyage to the Davis Strait grounds (Gale Huntington Research Library 1795; Hawes 1924:99; Starbuck 1878:110). Though not organized as such, several general divisions of goods can be differentiated including food and drinks, arms, shipbuilding tools, spare lumber, fasteners, vessel and rigging maintenance, whalecraft, cordage, galley wares, fishing equipment, and navigational aids. This list is the earliest such record identified in a survey of the archives of several repositories in New England. Most of the records surveyed indicated that, prior to 1765, materials intended for use on a particular whaling vessel were simply kept in the general account ledgers of local merchants or in those of vessel owners.

The increasing numbers of ships entering the fishery, the larger size of the vessels, and length of the journeys over the ensuing decades were all factors that led to an increase in the use of outfitting lists to better track goods shipped onboard. A review of the numbers and quality of these kinds of records located in whaling-related archives of New England indicates that a formalization of records occurred in the early 1800s. These records appear to have progressed from handwritten documents on unbound paper (NBWM 1800), to paper folded and bound booklets (NHA 1807), to ledgers devoted to keeping such records for a particular ship (NHA 1820), to pamphlets with preprinted pages listing necessary articles (NBWM 1832), and finally to booklets with decorative covers and preprinted pages listing a wide range of articles (Kirby 1860). As the nineteenth century progressed, the use of preprinted pamphlets and booklets appears to have become more common;

hundreds of such files dating from the 1840s to the 1870s are preserved in the Whaling Room of the New Bedford Free Public Library and are testament to this development. Though it appears that not all agents chose to utilize these preprinted lists in the early period, by the 1850s their use was commonplace.

Undoubtedly, the growth of the whaling industry and the number of printing presses available in coastal towns by the 1840s combined to make the use of these booklets even more prevalent. Most of the pamphlets included advertisements for ship chandleries, grocers, and makers of nautical instruments, and it is quite likely that they were printed by those businesses and distributed to the agents for free in the hope that agents would patronize the establishments. As the years of the golden age passed, the types of materials and number of categories included on these lists increased. Whaling historian Elmo Hohman indicates that by the mid-nineteenth century "the thousands of individual articles carried by a whaler, comprising more than one thousand *kinds* of goods, were classified under twenty-five different divisions" (1928:331). A complete listing of all of the materials in outfitting books of early to mid-nineteenth-century whaleships is outside the scope of this work. However, the complete text from one used for the ship *Condor* (1832) is shown in the appendix. A review of the many and diverse items on that list make it easy to see why one historian commented that a whaleship was really a "floating department store, carpenter shop, blacksmith shop, ship yard, and several other things all rolled into one" (Verrill 1916:67).

Regardless of the format used for keeping track of outfitting, agents and captains recorded a great deal more information in the books than simply checking items off a list. In most cases, the booklets tended to be used as a guide; the pages are often annotated with numbers beside specific items and any blank pages or empty spaces used for notes. Of course, keeping track of everything that made it aboard a ship would have been impossible, especially items brought by crew. Each general crew member was allowed a small amount of space for possessions, usually confined in a wooden sea chest or a canvas sack, in which they could keep personal supplies and souvenirs they might accumulate (Dodge 1976:57–58). For the most part, however, a thorough account was maintained, and the captain was aware of the items that could be found onboard.

Because the quantity of materials and provisions required so much space, a simple but effective system for storing it all was devised: encasing them in the same barrels that would later be used for whale oil storage. Though the term *cask* was generically used for all of the wooden containers shipped onboard vessels of the nineteenth century, they ranged greatly in size, and each had a particular name (Staniforth 1987:21). Casks were staved containers composed of three major parts: wooden staves, which were the numerous planks made to fit tightly on two sides with other staves to form the curved sides of the cask;

head pieces, the planks used to create the upper and lower ends of the container; and hoops, narrow strips of wood or iron placed around the finished cask to tightly bind the seams (Ross 1985:3–8). All casks were constructed of oak, which was highly regarded as being "flexible, strong and close-grained, it is impervious to liquid, hard-wearing and will not usually crack when heated" (Howard 1996:437). Though the British used at least five different types of oak for cask construction, American coopers preferred New England white oak (Howard 1996:437). Although wooden hoops were used for hundreds of years by whalers working in the cooler climes of the North Atlantic region, by the early nineteenth century the hoops used for binding casks were made solely of iron. Unlike their wooden predecessors, iron hoops could be driven farther down to tighten seams when needed and stop leakage. Historian Mark Howard suggests that this quality was a precondition for whaling moving into equatorial seas since warmer climates increased leakage (1996:436).

Before the work of outfitting was completed, the whaling agent had contracted the services of a neighboring cooperage to begin manufacturing the barrels that would be used onboard. The cooper began by taking measurements of the ship's hold to determine the exact number and sizes of wooden casks that would best fit the ship, and with this information, a gang of workers quickly engaged in crafting the enormous number of casks needed (Brown 1887:237–238). Casks were of utmost importance to a whaling cruise as they were the only reliable containers available in the early to mid-nineteenth century. They could be broken down and carried unassembled, and the wood expanded when liquid was added, which kept leakage to a minimum (Morton 1982:58). Whaleships carried thousands of casks, many of which were disassembled and stored in sets of staves and head pieces called "shooks" (Olly 2004:150). Several shooks were often stored within a single, large cask; when a whale was taken, these were broken out and easily reassembled by adding iron hoops to hold them together. Since casks were essential to industrial success, their construction and reassembly took a great deal of skill. Therefore, good coopers were always in high demand and among the best paid of the crew of early nineteenth-century whaleships.

When the whaleship was ready to leave the dock, it was loaded from keel to deck with casks—sometimes referred to as hogsheads—in a range of sizes (Doane 1987:74; Grant 1932:40). The most common way to stow casks was the "bilge and cantline" method, which involved placing the bottom layer of casks facing fore and aft next to one another, directly on the ship's ceiling (inner) planking and then loading subsequent levels—also facing fore and aft—such that the bilge, or widest part, of the cask lay in the cantline, or hollow, formed by the casks below them (Staniforth 1987:22). The largest casks formed the ground tier in the hold and served as ballast; they were filled with fresh water or salted meat to prevent shrinking or rotting (Doane 1987:74). Long, narrow casks known as "ryers" filled empty spaces and helped to secure those filled with articles such as spare sails, food, and clothing for sale to the crew (known as slops and kept in the slop chest), which were deliberately stacked on top of one another in tiers (Grant 1932:40). Over the course of a successful cruise when the reserves of shooks and stored provisions were depleted, any available cask would be used for oil storage; this included the bottom tier ballast casks, which were pumped out, swabbed dry, and then refilled with oil (Doane 1987:74; Morton 1982:58). This system for recycling casks proved to be very effective and typified the general attitude of owners and agents toward unnecessary waste or expense.

Ballasting

It was common onboard whaleships to use liquid-filled casks for ballasting a ship, but a small amount of standard ballast was also added to spaces that casks did not fill. *Ballast* is a blanket term used for any heavy materials placed low in a ship's hold to increase its stability and to regulate its trim by reducing the center of gravity (Desmond 1919:201; Steffy 1994:267). Frequently used ballasting materials included stone, iron, lead, bricks, and, in the case of whaleships, liquid. Iron bars known as "pigs" or "kentledge" were often used onboard sailing vessels, particularly by the British Navy, throughout the eighteenth and nineteenth centuries (Tuttle et al. 2010:42–43). The most common materials for this purpose, however, were river stones because they could be easily distributed where needed. Known as cobbles, these stones could be acquired for little or no expense, which makes it likely that they would have been favored by American whaleship owners and outfitters.

Archaeological Evidence for Ballasting

Approximately 30 ballast stones were recorded at the *Two Brothers* shipwreck. Of those examined, all were roughly 15 by 10 cm in size, varied in shape, and had the appearance of having been smoothed by river water. Although it is possible that a place of origin for these stones might be ascertained by a specialist (Stone 1993:14), ballast was commonly added and removed at the many ports visited by sailing vessels, which makes obtaining a conclusive determination from such studies difficult. Based on the small number of stones and the fact that they were found clustered mainly in a small pile in Section B of the site, it is likely that the ship was mainly ballasted using liquid and that cobbles were placed aft of the bottom tier casks to add weight. In contrast to this wrecked American whaler, the only type of ballast documented among the remains of the three British whaleships lost in PNMN between 1822 and

1837 was pig iron in standard lengths. This phenomenon may indicate a preference for metal over stone ballast among British owners.

Insurance and Certifications

Once the outfitting was complete and the equipment stowed, the lists were checked to assure that all the multifarious articles required for a whaling voyage were actually onboard before sailing (Hohman 1928:331). While this final inventory was conducted, the agent was busy tending to matters crucial to the ship's departure and safe passage. Such chores included finalizing the marine insurance for the vessel and its outfits, making sure that all crew members had signed contracts, and that all necessary documents were ready for the captain to take possession. Insurance was an increasingly important aspect of the whaling industry, and policies were generally underwritten by wealthy citizens or companies (Stein 1992:93). By 1820, marine insurance was readily available to all owners, and costs ranged greatly due to factors including vessel size and condition, estimated length of a proposed cruise, and the environmental conditions in the hunting regions (Hohman 1928:273–279). Regardless of the cost, most ships and their outfits were insured to some extent during the early to mid-nineteenth century since the potential for loss was so great.

The documents carried onboard a whaling vessel in this period were essentially the same as those required of all merchant vessels. Documentation resulted from legislative efforts by the US government to regulate and protect maritime commerce between 1776 and 1860 and the documents generated by those statutes were issued by the US Customs Service and US Consular Service (Stein 1992:12). The *Ship Register*, which provided official sanction from the federal government by acknowledging the ship as a representative of the United States was one of the most important documents (Allen 1973:160). Documents relating to the crew of the ship included *Articles of Agreement*, which were effectively contracts between owners and the crews of merchant vessels (Davis et al. 1997:87), or a *Whalemen's Shipping Paper*, a type of crew list unique to the whaling industry that included the specific conditions for the whaling voyage, the name of each crew member, and the amount of their lay (Sherman 1965:59–60; Stein 1992:154). Other documents in possession of a whaleship's captain included *Certificates of Clearance*, *Consular Certificates*, and *Bills of Health* (Davis et al. 1997:87; Sherman 1965; Stein 1992:154).

After the final checks were completed and the required documents were secured, the vessel was ready for departure. Since whaling was the main livelihood of many New England ports in the early to mid-nineteenth century, sailing day often brought great activity to the wharves as owners, their families, and the townspeople crowded the wharf to wish the whalers "greasy luck" (Grant 1932:2). Once the officers and crew were aboard and accounted for, "the anchor was weighed, the sails were sheeted home, and the whaler slipped passed the harbor lights—outward bound" (Verrill 1916:95). Thus began the next stage of the pelagic whaling system, the cruise operations stage.

Postcruise Stage

The postcruise stage of the pelagic whaling system was the least intensive part of the trip. It involved unloading cargo and preparing the vessel for refit, as well as settling accounts and formally registering any legal issues. The factors included in this stage could range from simple to complex depending on the success of the cruise and the conduct of the crew. Though detailed descriptions of all aspects of this stage are interesting in historical perspective, for the purpose of the discussion of how the remains of wrecked whaleships can be interpreted as maritime industrial workplaces, they are subsidiary.

Upon returning to port, the work of unloading the cargo and clearing the ship was carried out by local workers while the ship's crew was released from duty. Before the whalers could be paid, it was first necessary to determine the amount of profit, if any, that the cruise generated. This process sometimes took as many as three weeks, and the whalers were obliged to remain in port until it was complete (Davis 1874:394–395). The process began with the ship being "broken out" by stevedores, who were employed by the outfitter and who hoisted the casks from the hold (Ashley 1926:98). Once the ship was emptied of all cargo, equipment, and any remaining provisions that might still have value, it was thoroughly inventoried and assessed for possible reuse on later voyages. All whaling tools and equipment were then placed in the lofts of buildings belonging to the owners, and the oil was either sent to a warehouse or stored on the wharf (Figure 5.7), often with seaweed piled onto the casks to keep them from drying out and leaking (Brown 1887:230).

Once complete, the final inventories were then taken to a countinghouse, where clerks settled all accounts before determining profits and the crew's wages. As stated in their contracts, fees were deducted from earnings for services including sales commissions, insurance, outfitting, pilotage, wharfage, shipping fees, and cooperage (Dolin 2007:270; Hohman 1926:645). When the painstaking process of accounting was complete, the lays of all members of the crew were known. Before payments were made, however, subtractions were made for all deductions, advances, and credits that had been recorded in the ship's log and other ledgers or notebooks kept by the captain (Hohman 1926:646). As Hohman notes, "the most significant of these deductions comprised the slop-chest account, containing a notation of all the supplies which any man had purchased on credit from the ship's store, or slop-chest; the amounts of cash which had been advanced to him, mainly to provide spending-money while in port; and the interest charges, calculated at

Figure 5.7. Mid-nineteenth-century photograph of casks of whale oil stored on a dock at the New Bedford waterfront. (From State Street Trust and Walton Advertising and Printing 1915.)

generous if not exorbitant rates, on the sums advanced" (Hohman 1926:647). Once these charges were tabulated and deducted, the crews finally received their wages, which were generally paid in cash or in a bill of exchange that would be converted to cash (Davis et al. 1997:157).

Conclusion

American pelagic whaling operated under a strictly regulated industrial system from the mid-eighteenth to the early twentieth centuries. This proved to be highly successful due to the standardization of not only the physical process involved in pursuing, capturing, and processing whales but also the decision-making processes and preparations prior to departure and the tasks involved in the conclusion of the business and readying the ship for its next voyage. As such, the focus of the preceding discussion on the beginning (precruise stage) and ending stage (postcruise stage) of a whaling cruise provides insight into the many steps involved in ensuring the success of a voyage. While these stages bookend the system and provide the necessary elements for the maritime industrial workplaces onboard the ships to come to life, each was just as important as the others for the success of a whaling trip.

6

Roving Production Platforms in the Pacific

Although the three parts of the American pelagic whaling system were all equally important, the cruise operations stage has long captured the imagination of the public. The activities that took place during this stage of the cruise are most often described in popular literature on whaling. Though actions such as scanning the horizon from the lookout or going into battle with a great whale were the central aim of the cruise, they were only a couple aspects of the operations onboard an early to mid-nineteenth-century whaleship. To better understand the daily routines, this chapter discusses cruise operations and is separated into sections relating to working and recreation periods, as well as matters pertaining to meals and the general health of the crew. The methods for capturing and processing whales are also described separately, since during these periods, the regular rotation of daily chores and downtime was suspended. To integrate a discussion of the significance of many of the whaling-specific artifacts identified at the *Two Brothers* and *Parker* shipwreck sites, historical descriptions of daily life and the industrial processes employed are followed by discussions of related artifacts found in archaeological contexts.

Organization and Fitting Ship

With the land beyond the horizon, the whaleship became a world of its own (Allen 1973:162). The crew was organized immediately so that work could commence. This process began with the division of the crew into two watches, which were short working periods designed to ensure that part of the crew was always alert. Designated as the starboard and larboard (port) watches and overseen by the second and third mate respectively, one of the watches (e.g., half the crew) was always on deck during the day (Brown 1887:229). The term *watch* was used to refer to both the amount of time and the crew assigned to it (Dana 1842:129). The organization of labor into watches was common practice on all merchant vessels of the early nineteenth century.

Watches were closely regulated and while making the passage to the whaling grounds, each watch worked four-hour shifts throughout the day and night (Doane 1987:275; Olmsted 1841:47). Between four and eight o'clock in the evening, these were broken into "dogwatches" of two hours to keep the rotation even and to allow the crews to eat dinner and relax (Brown 1887:229; Scammon 1874:222; Shapiro 1959:25). To ensure that the rigid schedule was maintained, two objects were critical: the ship's bell and a sandglass. In his marine dictionary, J. H. Röding indicates the relationship between the two by stating that "at each change of watch, namely after four hours have elapsed and the sand in the half-hourglass has run out eight times, the bell is rung to awaken the watch below to come up" (Röding 1793; Wede 1972:4). As demonstrated by the New Bedford Whaling Museum's half-sized model of the mid-nineteenth-century whaleship *Lagoda*, the bell and its associated assembly were situated in a framework mounted atop the samson post at the front of the windlass.

With the organization of the crew complete and the roster set, it was customary for the master to address the crew on the first full day at sea (Kugler 1971:5; Mawer 1999:64). The speech delivered by the captain was intended to define the expectations of the hands and explain that failure to meet them would result in disciplinary actions (Brown 1887:230; Verrill 1916:80–81). The gist of this relatively brief message varied little among whaling captains, who performed it without the aid of notes (Browne 1850:35–37; Dulles 1933:109–110; Grant 1932:6; Mawer 1999:64). One exception to the rule was that of Captain Edward S. Davoll, a whaleship master during the late 1840s and 1850s who kept a small notebook containing a written script for this purpose (Kugler 1971:6). Published in 1981 as a monograph titled *The Captain's Specific Orders*, the text provides a rare glimpse of the expectations for the officers and crew aboard a mid-nineteenth-century whaleship. It is divided into the following sections: "To the Foremast Hands, Cook of the Vessel, Steward, Boatsteerers, Officers," "To the Mate," and "To the Second Mate" (Davoll 1981). Each of these sections addressed things like duties of each position, rules to be followed, respect for others onboard, particular language to be used during operations, maintenance schedules, punishments, and particular physical boundaries to be maintained while on deck by the different ranks (Davoll 1981). Though not specifically stated, the data contained within Davoll's notes was likely gleaned from his experiences while working up the professional ladder—or as whalers referred to it, "through the hawsehole"—from the age of eighteen and likely represents many of the same instructions that were given to him by previous captains (Whipple

1979:64). It might be noted that Davoll did not include position descriptions for the cooper, carpenter, or blacksmith; this is presumably due to the fact they were highly skilled craftworkers who knew their jobs and needed little instruction.

The first few days at sea often provided an opportunity for the crew to get their sea legs, since many of the green hands needed time to overcome the debilitating seasickness often associated with a first trip on the ocean, and some of the older sailors desired the same to sober up from their last round of "parting drinks" (Hawes 1924:240; Scammon 1874:223). The amount of acclimation time afforded the crew depended entirely on the disposition of the officers; some were generous and recognized the issues as temporary, while others were more callous and demanded that the hands be put to work immediately (Dulles 1933:107).

Regardless of the period allowed for the crew to adjust, the first task needing attention was that of "fitting ship" (Scammon 1874:223). Overseen by the first mate, fitting ship involved not only the usual work aboard any sailing ship of the period, such as tending to sails or scrubbing the decks. It also meant readying the ship for whaling operations by "breaking out" the casks containing the gear needed for the different processes of pelagic whaling (Davoll 1981:18; Dulles 1933:110; Scammon 1874:223). Among the many chores were reorganizing supplies; making brooms; outfitting the boats with the myriad supplies and whaling implements they required (see the appendix); carefully coiling lines into tubs for attaching to harpoons; almost constantly turning the grindstone to ensure that all of the harpoons and lances, as well as the instruments associated with cutting in, were razor sharp; and setting up the tryworks, skimmers, ladles, and all of the other tools employed in rending blubber into oil (Dulles 1933:110; Grant 1932:42–44; Hegarty 1960:37; Perkins 1854:25; Scammon 1874:223; Shapiro 1959:25).

On Watch

Once the officers were content that fitting ship was complete and the workplace was shipshape, the watches settled into the routine chores expected of them. The schedule of work onboard a whaleship revolved around several regimented activities that were performed without fail. These ranged from tasks of utmost importance of the conduct and maintenance of the ship and the industrial success of the cruise to more mundane chores like weaving mats for chafing gear (Brown 1887:229; WPA 1938:14) or cleaning the decks using scrub brooms made from pickled blocks of seasoned oak (Perkins 1854:29). Other activities that were attended to as part of the watch included patching, painting, caulking, pumping the bilge, and—once oil had been procured—"wetting the hold" or keeping casks wet to prevent leakage (Ashley 1926:98; Brown

1887:289; Dickson 2011:3; Doane 1987:74). Thus, maintenance of the ship and cargo occupied a large part of a sailor's working life, especially during the four- to six-month outbound journey to the Pacific, when whales were not expected (Delano 1846:80; Dulles 1933:112; Morton 1982:87).

Among the most important duties during the first part of the cruise was training the green hands onboard. Obviously, this involved teaching them all manner of things associated with working on any sailing ship of the period such as handling sails and lines, tying nautical knots, and splicing. Whaleship crews, however, also needed to be able to launch the whaleboats in pursuit of prey at a moment's notice, and seemingly endless drills helped them to fully understand that strict routine (Stackpole 1967:37). Training included instruction in the exact methods for lowering the boats from the davits and entering them, handling the sail and oars, techniques for approaching and killing whales, communication with the ship, attaching the carcass to the ship, and recradling the boats. To assist with these exercises, an extra spar was often lowered into the water and used as a dummy whale for the crews to learn to maneuver around. Drills continued throughout the outbound voyage, especially when the ship was becalmed, until discipline among the crew was established (Browne 1850:38; Payne n.d.:2; Verrill 1916:97). The value of this training cannot be understated; whaling was an extremely dangerous activity that demanded attention and proficiency on the part of whaleboat crews. When considering the range of skills to be mastered during this relatively short period, the training of green hands on a first cruise can be seen as an unofficial period of apprenticeship (Hohman 1926:658).

Also of utmost importance among the routine daily activities onboard a whaleship was manning the lookout. Since the obvious objective of the cruise was to take whales, no time was wasted in positioning crew high in the masts for this duty. Each day at dawn, an officer or boatsteerer went aloft to the head of the mainmast, and one or two hands ascended to the fore-topgallant crosstrees to scan the horizon for whales (Payne n.d.:2; Weiss et al. 1974:82). Around the middle of the nineteenth century, small platforms were constructed at the heads of the masts that included a double-ringed iron railing, referred to as "the hoops" (Stackpole 1967:38). This innovation increased safety and reduced the risk of falling as before its introduction crew members on lookout duty simply held on to the spars and mast (Figure 6.1). Sharp eyes were of greatest importance for this duty, so the lookout crews were relieved every two hours from dawn to sunset (Browne 1850:192; Macy 1835:223). As one of the main responsibilities involved in the watch, officers of whaleships kept a close eye on the crew members at the masthead to ensure that they were alert and that no prey was missed.

Figure 6.1. Illustration of a whaler aloft and signaling that a whale has been spotted. (From Davis 1874.)

The rotation of the lookouts coincided with two-hour watches at the ship's steering wheel (Browne 1850:192; Perkins 1854:30; Scammon 1874:229). Standing watch at the wheel involved watching the sails and rigging, listening for calls from the lookouts, and keeping track of the needle compass mounted just forward of the helm (Higgins 1927:22). Maintaining the course established by the captain while transiting to a targeted whaling ground was an important duty of those watching the helm, particularly once the ship entered the Pacific. Though innovations in the steering apparatus were common on merchant ships after the 1860s, American whaleships always used a type known as a "travelling wheel" (Grant 1932:28; Higgins 1927:22). This system involved the head of the rudder post protruding through the deck with a tiller connected; attached to the tiller was a helm consisting of a wheel and drum. The mechanism worked through a system of rigging that allowed it to move easily. Ropes were wrapped multiple times around the drum, then looped through blocks on the tiller and on to another set of blocks attached to the bulwarks on each side of the ship (Grant 1932:28; Higgins 1927:22). This projecting design, referred to by whalers as a shincracker, required the helmsman to walk back and forth across the deck when the wheel was turned, which in turn kept them busy and alert (Grant 1932:28; Leavitt 1973:135).

Above deck, the shincracker steering system was composed mainly of wooden and fiber components. The version of this assembly used on early nineteenth-century whaleships had few metal components aside from some fasteners, a rod used for the drum axle, small sheave pins in the steering blocks, and eyebolts to which the blocks connected (Higgins 1927:22). As evidenced on both the half-model of *Lagoda* and the 1841 whaleship *Charles W. Morgan*, by the middle of the nineteenth century, more metal parts were added including a strap and fasteners used to better secure the tiller to the head of the rudder post and a brass frame fitted over its front and rigged to the steering blocks.

The other main component of the steering system was the rudder. Affixed directly to the tiller and extending into the water to the depth of the keel, this large timber blade was hinged against the sternpost using pintles and gudgeons which allowed it to pivot and control the ship's direction while underway (Davis 1918:32; Steffy 1994:278). The rudder was composed of vertical planks that acted as a paddle, and the forward most of them—known as the rudderpost—projected up through the deck. The front edge of the rudderpost was notched to accept the four evenly spaced pintles, which were straps connected to the rudder that had downward projecting pins on their faces. These pins were used to slot into a corresponding hole on the faces of gudgeons, which were straps that fastened to the sternpost (Nash 2009:135). As with the rest of the hull, the rudder was protected from marine borer infestation by adding first a horizontally lain layer of 1.5-cm white pine and then a layer of copper or Muntz sheathing running from its bottom to 30 cm above the ship's load waterline (Hegarty 1964:49–50).

Because rudders were not fastened securely to the ship, a removable block known as a "wood lock" was often added below one of the pintles to keep the rudder from floating off its hinges (Luce 1863:91; Stone 1993:59). Heavy ropes or chains, both referred to as the "rudder pendant," were connected to eyes mounted near the top of the rudder and used to tether it to the stern. With the rudder attached in this way, it was possible to rehang or "ship" it while at sea if it became dislodged (Luce 1863:91; Stone 1993:59). In some instances, however, weather conditions were so extreme that pendants broke, and rudders were lost altogether, which rendered ships uncontrollable and often resulted in losses.

Directly related to steering the ship was navigation, which was among the many daily activities of the captain but usually carried out by other officers. As noted previously, captains of early nineteenth-century whaleships were required to be expert navigators. The search for new whaling grounds led ships ever farther into unexplored regions where they often passed previously unknown islands, reefs, and shoals. Since it was not until 1828 that the US government began to consider mounting a mapping expedition of the Pacific, new geographic features encountered by whaleships were ascribed a name—often that of the ship or captain recording them—and their approximate latitude and longitude were charted (Kugler 1971:25). To prevent tragedy, whaling captains shared these locations, and this practice gradually led to an expanding, communal knowledge of a particular region. Thus, of the many duties attended by the captain and officers while the ship was underway, maintaining position was of utmost importance.

Whenever whaleships encountered areas where the depth of water and the character of the bottom were not well-known, measurements were taken using a weighted line called a sounding lead (Luce 1863:345). Sounding leads were standard components of the navigational equipment for all ships of the period and consisted of a line of varying length, which was attached to an elongated piece of lead that tapered out from top to bottom. Depth was obtained by counting knots tied in the line that indicated a specific number of fathoms (1 fathom = 1.82 m). Bottom composition was ascertained by viewing the "arming," which was a lump of tallow placed in an indentation on the bottom of the lead used to capture a sample of sediment when it hit the seabed (Raper 1840:91). To use this tool, the "leadsman" threw it ahead of the ship while it was underway; the lead sank as the ship progressed, and the depth was taken when the line was perpendicular (Luce 1863:3). When a captain considered the ship to be nearing a geographical feature, soundings were taken so that the course could be changed to avoid shallow water and potential grounding.

Two types and sizes of leads were used for sounding in the nineteenth century. The smaller of the two, known as a hand-lead, weighed 7 to 14 pounds (3.2 to 6.4 kilograms [kg]) and was rigged with a line of 20 to 30 fathoms (36 to 55 m) that was coiled by hand. The larger, known as a deep-sea lead, weighed between 28 and 100 pounds (12.7 and 45.3 kg) and was rigged with 80 or more fathoms (146 m) of line often wound on a reel (Luce 1863:3; Raper 1840:91). Historic records show that both types were included among the many other articles carried onboard whaleships by at least the early 1830s (NBFPL 1845; NBWM 1832). While it is certain that sounding leads were used as early as 1765, the records give no indication of their type or weight (Gale Huntington Research Library 1765).

Once ships reached the deep water of the various Pacific grounds, they drifted along under shortened sail in search of whales. To ensure that they avoided any known hazards, it was important to record progress by observing drift, which was done using a log line and log glass (Fonda 1969:8; Freeman 1951:240). In the early 1800s, relative longitudinal positions were ascertained by taking lunar observations. This method involved using a quadrant to "observe the moon's angular distance from the sun or a suitable star" and then comparing the local time with that of Greenwich Time, which was derived from a table predicting the moon's position (Skelton 1954:109–113; Thrower 1967:35). Though for decades whaling captains used lunar observations with success, their reliability was subject to clear skies, so extended periods of poor weather could result in inaccurate positioning and possible shipwreck (Philbrick 2001:208).

Since errors in longitudinal observations essentially resulted in lost revenue by requiring more time to reach a targeted whaling ground and complicated the identification of reported features, by the early 1820s many whaling captains began using marine chronometers (Boggs 1938:180; Littlefield 1906:13). Originally perfected by English clockmaker James Harrison in 1759, the chronometer revolutionized navigation through the ability to "accurately record the time of day at a point of origin for comparison with local time, wherever the navigator happened to be" (Baker 1984:222; Thrower 1967:35). Essentially portable precision clocks, marine chronometers proved to be easier than taking lunar observations (Bartky 2000:12). Thus, as their accuracy increased and price decreased, they were soon among the most important articles of the ship's outfit despite their expense (Littlefield 1906:13; Songini 2007:330). Such was their value to captains that, by the early 1830s, they were already included among the list of navigation tools in preprinted outfitting books (NBWM 1832).

The marine chronometers used on early nineteenth-century whaleships consisted of a timepiece that was powered by a continuous spring and encased in a housing with a glass lens on the face (Porthouse 1848:6). Sometimes referred to as a "box chronometer" since the entire mechanism was protected by a hard wooden box, these sensitive instruments were suspended in brass gimbals affixed to the interior of the box; this allowed them to move freely with the motion of the ship (Porthouse 1848:6–7).

The compass was another crucial piece of navigational equipment and was used to determine direction and maintain course. Although the actual number that might be onboard a particular whaleship of the early nineteenth century varied, they were generally outfitted with three types of gimbaled compasses.

These included the large compass, which was used in the steering of the vessel and mounted inside the gabled, glass-windowed skylight just forward of the helm (Higgins 1927:22); the "telltale" compass, which was designed to be read upside down and would have been mounted over the captain's bunk to allow him to check the heading without getting up (Hegarty 1964:78); and the boat compass, which was a small compass housed in a wooden box and mounted on each whaleboat for use in navigating back to the ship when necessary (Ansel 1983:58; Ronnberg 1985:28–29).

Among the navigational aids in the possession of all whaleship captains was a copy of Nathaniel Bowditch's *The New American Practical Navigator*. First published in 1800, this work was "intended to acquaint the reader with the whole field of tasks of the ship's officer" (Fairburn 1945:1067). Though other similar guides were available, Bowditch's text was considered superior since he corrected errors in those published before. Adopted as the "Seaman's Bible," a copy became part of the necessary equipment of every ship's officer, as well as those aspiring to take up an officer's berth (Fairburn 1945:1067). *The New American Practical Navigator* is said to have given impetus to the transition from sailing by dead reckoning to the use of scientific celestial navigation, and by the 1820s, multiple copies of this book were found onboard most ships (Fairburn 1945:1068). On some whaleships of the period, the captain and mates encouraged eager members of the crew to study navigation and provided access to the books, instruments, and instruction in their use (Calkin 1953:23).

Evidence of the significance that whalers placed on Bowditch's navigation guide is provided in the accounts of survivors of the *Two Brothers* shipwreck. As discussed in chapter 4, the wrecking occurred with little time to abandon ship before it went to pieces in heavy seas. Though the crew left the ship with "no water nor provisions except two small pigs that washed into one of the boats," one of the mates made certain to grab a compass, one quadrant, and two "Practical Navigators" (Gardner 1823). From this it is presumed that the officers realized that once they were able to secure themselves in the whaleboats, their survival could rest in their ability to make an open sea voyage. Thus, having a copy of *The New American Practical Navigator* and navigation instruments was of greatest importance since it could increase their chances for success.

Along with advances in scientific navigation and more affordable instruments, mid-nineteenth century whalers benefited greatly from the US government's eventual efforts to chart the Pacific whaling grounds and to understand ocean currents and winds. Among the most important of these were the published results of the official exploration expedition led by USN Lieutenant Charles Wilkes (1842), which included a chapter devoted to the (then) popular whaling grounds. Also of significance were the works of Lieutenant Matthew F. Maury, whose 1850s maps helped to predict the best grounds to fish by season (Kugler 1971:25). Interestingly, both authors relied heavily on the experience of whaling captains who provided them navigational, geographic, and physical data. In Maury's case, data was recovered by studying logbooks and by enlisting captains to record observations at sea three times daily using a standard format. In exchange, he produced the now famous *Whale Chart*, which illustrated seasonal whale migrations between grounds and greatly benefited the whale fishery (Fairburn 1945:1063; Freeman 1951:240–241; Jones 1986:695; Kugler 1971:25; Mawer 1999:260). Along with the aforementioned navigational tools, these publications assisted in the daily navigation of the whaleships and, in many cases, helped to avert disaster.

Daily observations were taken from the deck of nineteenth-century ships and generally recorded on a "slate-log" or a "scrap logbook" (Wilson Barker and Allingham 1896:41). Listed as standard items in outfitting books from at least the early 1830s, writing slates were often inscribed with columns for inserting information such as time, weather, distance traveled, and comments pertaining to the sailing of the ship (Stanbury 2010:15; Young 1890:25).

Keeping whaling implements sharp was among the most important of the duties assigned to average seafarers. This task cannot be understated since not only did dull harpoons and lances greatly decrease the probability of capturing a whale by not fastening properly, ineffective attempts at doing so often led to boats being destroyed by the jaws or flukes of an infuriated whale (Verrill 1916). The outcomes of either of those scenarios were unfavorable since they ultimately resulted in the loss of profit. Thus, what might seem to be an unimportant component of a carpenter's toolkit was in fact crucial to the success of the cruise and must be considered among the most valuable tools of pelagic whaling.

The grindstones used onboard early nineteenth-century whaleships were composite tools generally consisting of a wooden frame, a stone disc, and an iron crank. The frame was approximately 90 cm tall, rectangular in shape, with an open center and a beveled area on each side of the upper rim. The iron grindstone crank was inserted through a hole in the center of the disc and placed in the bevels to create an axle on which the stone revolved (Dow 1925). The grindstone wheel was often lubricated with water to keep it cool and free of residues (Light 2007:120). Though grindstones were portable objects, they were most often situated on the main deck aft of the carpenter's bench (Whipple 1979:92–93). Sharpening was undertaken by two or three crew members, one who held the tool to the stone and the other(s) turned the crank and kept it wet (Grant 1932:42).

For most of a whaleship's crew, the day began with breakfast for all hands at sunrise and ended each day at around 4:00 in the afternoon after the decks were scrubbed and the tools put away (Browne 1850:38; Perkins 1854:30; Scammon 1874:229; WPA 1938:14). The masthead lookouts maintained their positions until sunset; at that time the ship was slowed by having its sails shortened, and the watch at the helm was set for the night (Perkins 1854:30; Scammon 1874:229). In the period between work ending and sunset, the two-hour dogwatches began. With dinner served at 5:00, the shortened work period allowed each watch the opportunity to eat and rest (WPA 1938:14). After the completion of daily operations, information recorded on the log-slate was transferred to the main ship's log along with comments about daily activities such as the conduct of the ship and any whales sighted or taken (Wilson Barker and Allingham 1896:41; Young 1890:25). As the official record of the voyage, logbooks were kept by the first mate who often added a phrase such as "thus ends this day's work" or "so ends this day" to indicate that nothing more would be accomplished (Calkin 1953:23; Dodge 1882:13; Sherman 1965:32; Verrill 1916:163; WPA 1938:29).

Archaeological Evidence for On Watch Activities

A number of artifacts related with activities undertaken during the watch were noted at the *Two Bothers* and *Parker* shipwreck sties. Since the remains of the whaleship *Parker* at Kure Atoll are mostly associated with the bow of the ship, it is unsurprising that a bell was also documented in its assemblage. Recovered for possible ship identification as well as exhibition, conservation of this artifact revealed a composition of almost solid copper (98%). And while the iron clapper used for ringing and iron yoke (encased in wood) from which the bell was suspended were present, the name of the ship was not inscribed on its surface (Fox 2010). The fact that this bell was plain and unadorned, however, might be taken as more evidence of its function as a whaleship. Due to the general thriftiness of owners and agents, they were more likely to purchase a generic bell off the shelf at a local ship chandlery rather than justifying the expense of having one cast specifically with the ship's name (Wede 1972:44).

Another material component possibly associated with watchkeeping duty was identified among the remains of *Two Brothers* at French Frigate Shoals. A single shard of thin, clear glass that appeared to conform to the shape of the globular end portions of a sandglass was identified in Section B. It is important to note, however, the small size of this artifact prevents it from being positively associated with the watch since sandglasses of various sizes were used onboard ships for different timing purposes. Factors affecting the amount of time that a sandglass measured included its size and the fill material (Ford et al. 2008:146–148). To determine the particular use for a sandglass would require a shard large enough to indicate overall size; the scant dimensions of the artifact recorded at the *Two Brothers* shipwreck site gives no indication of a possible height or diameter for the sandglass. Thus, while it is possible that this artifact was used in conjunction with timing of watch periods, no conclusive determination can be made.

Because of its importance to operations and its dense physical composition, it is no surprise that a grindstone was identified among the remains of the wrecked whaleship *Two Brothers*. Resting in a pocket on the reef top along with a small number of fasteners and other iron objects, this discoidal stone has a square hole in its center and showed no evidence of markings or whether it has been used previously (Figure 6.2). As with most important industrial objects, redundancy was a consideration when outfitting a whaleship; as such, at least one spare grindstone was included among the carpenter's tools. Made of durable sandstone, grindstones were among the equipment that might be reused for multiple whaling cruises. Found in the suspected stern portion of the wreck site (Section B), it is possible that this stone was not in active use at the time of wrecking and was instead a spare that was stowed away. A similar grindstone was recorded on the wreck of the British whaleship *Pearl*, which was lost in 1822 at Pearl and Hermes Reef in the NWHI (Van Tilburg 2005).

Figure 6.2. Grindstone documented in a pocket on the reef top at Section B of the *Two Brothers* shipwreck site. Note numerous unidentifiable concreted iron objects nearby. (Courtesy of Papahanaumokuakea Marine National Monument.)

Four pintles found situated in the area between the back reef and the main concentration of the *Parker* shipwreck site are evidence of the ship's rudder and, therefore, associated with keeping the helm. As was standard for the nineteenth century, these objects were composed of copper (or a cuprous alloy), and their number and varying sizes indicate that they represent a complete set. Based on the area in which they are found and the fact that corresponding gudgeons were not identified near them, it is presumed that the rudder was unhinged from the ship at some point during the wrecking event. Since most of the fasteners remain intact or mostly intact, it is likely that the rudder timbers deteriorated in situ. Though the pintles identified at the *Parker* site are in excellent condition, no visible markings are present to identify the foundry of manufacture.

A small, octagonal-shaped sounding lead was found concreted to the reef in Section B of the *Two Brothers* shipwreck site. Clearly exhibiting an attachment eye at the top end and an indentation on its bottom, its size suggests that it is a hand-lead (Figure 6.3). As noted, sounding leads were integral to understanding the depth and seabed in which a vessel was operating. Although sounding leads recovered from other shipwreck sites of the period include numerical marks, the one found at *Two Brothers* was recorded in situ, and no markings were visible due to marine growth.

A wedge-shaped shard of clear glass identified at the wreck of *Two Brothers* may be part of a lens from a marine chronometer. This 3-mm-thick shard appears to represent approximately one-quarter of a round lens that was 14 cm in diameter. Recorded in Section B of the wreck site, it rests near several other artifacts associated with the navigation of the ship. Although the size and shape of this shard closely matches those of box chronometer lenses from the period, its identification is questionable as accounts indicate that Captain Pollard was forced to rely on dead reckoning due to poor weather, which might indicate that he was sailing without a chronometer in 1823 (Nickerson 1876; Philbrick 2001:208).

The wedge-shaped shard of clear glass found at *Two Brothers* could, however, be a portion of the lens from a small compass. The lens of the large compass is estimated to have averaged 25 cm in diameter, while the average diameters of the lenses of the telltale and the boat compass are estimated at 15 cm in diameter. Based on the dimensions of the wedge-shaped shard documented at the *Two Brothers* site, it is undoubtedly a portion of a lens 14 cm in diameter when it was whole. Since this artifact was recorded in the stern portion of ship, it likely that it is part of the telltale compass positioned in the captain's quarters or a spare boat compass.

Finally, a piece of slate found on the *Two Brothers* shipwreck could be a portion of a log-slate. Because it was recorded in situ, the marine growth covering

Figure 6.3. "Hand lead"–style sounding lead documented at Section B of the *Two Brothers* shipwreck site. (Courtesy of Papahanaumokuakea Marine National Monument.)

the object prevented the detection of markings on either side. Though this artifact was in the presumed stern section of the vessel along with other navigational equipment, it is unknown whether it was in active use at the time of wrecking or if it was stored. A similar piece of a log-slate was identified during the investigation of the whaleship *Lady Lyttleton*, lost on the southern coast of western Australia in 1867 (Stanbury 2010:15; Vosmer and Wright 1991:22).

Watch Suspended: Pelagic Whaling in Process

Aside from the daily activities associated with time spent both on and off duty, there was the constant prospect of whales being sighted. Whale fishing was the central focus of the cruise, and when it occurred, the workplace changed; the rotation of the watches was postponed, all hands were immediately required on deck, and the ship bustled with activity in all quarters. Since whale hunting was an opportunistic pursuit, a missed whale effectively meant missed profit; thus, when whales were seen, all other shipboard operations were suspended in preparation for pursuit. Capturing a whale and recovering the valuable commercial products from its carcass was a cyclical process involving a series of steps completed in order and designed to always keep crews and equipment ready. For the most part, these steps did not start and stop when the previous one was complete; instead, they were fluid and could occur simultaneously if necessary (Brown 1887:286). As stated, this operational formula developed with the installation of the tryworks onboard whaleships in the mid-eighteenth century and evolved into a structure that saw specific tasks carried out by specific crew members using specific equipment. The following discussion provides an overview of the process as a whole and defines the five interrelated steps required for success.

Raising Whales and Capture

Daytime lookouts stationed at the tops of the masts from the outset of a whaling voyage constantly scanned the horizon looking for the low spouts of vapor that indicated the presence of a sperm whale (Leavitt 1973:16; Shapiro 1959:32; Verrill 1916:103). When whales were sighted or "raised" (Scammon 1874:229), the lookout announced their presence with the call of "there she blows" or some derivative of that phrase (Brown 1887:257; Browne 1850:115; Cheever 1850:52; Davis 1874:29; Delano 1846:28; Dodge 1882:12; Olmsted 1841:55; Scammon 1874:230). Upon hearing this call, the captain demanded the direction of the spout and often went aloft to survey the scene; there he could better determine the whale's distance from the ship and the chances of capture (Leavitt 1973:16; WPA 1938:21). Meanwhile, the hands in the forecastle swarmed through the companionway to gather with the rest of the crew on deck and await orders (Brown 1887:258). While the captain deliberated, the boat crews prepared the boats to be launched; if the whale was to be pursued, the captain chose the number of boats to engage (Leavitt 1973:16; Weiss et al. 1974:82). If the spout was far away and sea conditions permitted, the sails were trimmed, and the helmsperson steered the ship to position it to windward of the whale since approaching from downwind was always faster and easier (Allen 1973:171; Hohman 1928:156).

Since sperm whales possess a keen sense of hearing, it was necessary to keep the ship at least a mile distant; at that point the ship was "hove-to," and the call was made to "lower away" (Hegarty 1960:39; Macy 1835:225; Verrill 1916:103; WPA 1938:21). The mate and the boatsteerer boarded the whaleboat and then secured the tubs of whale line; the gripes were then freed, and the mate ordered the crew to hoist the weight of the boat to allow the cranes to be swung clear (Leavitt 1973:16; Stackpole 1967:39). The boats were lowered straight down, and once on the water, the rest of the boat crew slid down the lines or scrambled over the side of the ship to take their places. Once set, the boat was unhooked and pushed away from the ship (Allen 1973:172; Grant 1932:58; Weiss et al. 1974:82). Depending on the distance to the whale, either the boat's mast was raised and sails set or the oars were manned and the crew "pulled" in its direction. In the meantime, the boatsteerer prepared the first harpoon for action by attaching or "bending" the whale line to it and then wrapping a length of it around the boat's loggerhead to ensure control once the harpoon was fastened to the whale (Doane 1987:68; Grant 1932:60).

With the captain onboard one of the whaleboats, the group of hands left onboard (generally the cooper and cook) acted as shipkeepers and looked after the vessel (Howard 1996:441; Shapiro 1959:33). As the boats moved farther away, they communicated with it through a variety of signal flags and shapes flown high from the masthead. Signals were used to direct the boats to the location of their prey, inform the boat crews whether it had sounded or been killed or to tell them to return to the ship. Since the signals were particular to each ship, they were especially useful when a large fleet was cruising on the same ground (Scammon 1874:229; Verrill 1916:113–114). Among the tools that shipkeepers used to communicate with the whaleboats was a "masthead waif." Sometimes called a yonder by British whalers, this tool was composed of a canvas covered hoop fastened at one end of a 1.83 to 2.44 m long pole and used to indicate the direction of whales when the boats were long distances from the ship (Delano 1846:76; Scammon 1874:230; Verrill 1916:113).

As the whaleboat made progress over the water, the harpooner kept an eye on the masthead and conveyed directions to the mate. Positioned in the stern, the mate maneuvered the boat toward the whale using the long steering oar

(Brown 1887:259). When the boats were within 500 m of a whale, it was necessary to be as quiet as possible to keep them from being "gallied" or alarmed (Brown 1887:259; Leavitt 1973:126; Olmsted 1841:57). Thus, if sailing, the mast and sails were lowered, and if rowing, the boat crew "peaked" their long oars (i.e., took them out of the water and rested upright in slots in the gunnels) and took up smaller paddles (Cheever 1850:53; Davis 1874:157; Doane 1987:68; Dodge 1882:12; Leavitt 1973:21; Verrill 1916:104). Up to that point, the boatsteerer had assisted with propulsion, but he now turned his attention to preparing the craft by removing the wooden sheaths from the harpoons and lances and readying the whale line (Weiss et al. 1974:84).

The specialized implements designed for and used in capturing whales were known as "whalecraft" (Davis et al. 1987:26; Lytle 1984). Along with the heavy manila lines, harpoons and hand lances were among the most important pieces of whalecraft. In their basic form, these tools used for attacking and killing whales changed little from those used in the early days of the Atlantic whale fisheries (Kugler 1971:21; Spence 1980:114). While the technique for employing them also remained essentially the same, variations to their form introduced during the early to mid-nineteenth century greatly increased industrial productivity. Thus, the changes to whalecraft are identified as among the major technological innovations experienced by the fishery (Davis et al. 1987:60).

Though none of these tools was more essential than the others, harpoons were of primary importance since without them the boats could not fasten to the whale. Referred to in American parlance as "irons" (Kugler 1971:21; Leavitt 1973:125), the harpoons used on early nineteenth-century American whaleships featured a flat, sharpened head welded to an iron shank that tapered out at the opposite end to create a socket. To get the sharpest possible edge, the metal used in the manufacture of harpoons was of the finest quality (Harwood 1935:158). As noted by historian John Spears, "well-worn horseshoes and horseshoe nails were much used in forging these special weapons, and razor steel was used in making the cutting edges and points" (Spears 1910:206–207). The implement was completed when a 1.82-m-long wooden shaft—called an iron pole—was affixed into the socket and a short piece of rope was secured to the shank (Olmsted 1841:20). Called the "iron strap," this piece of line had an eye spliced into one end and was used for "bending," or tying into the whale line, when the harpoon was ready for use (Brown 1887:250; Doane 1987:67; Spears 1910:206).

As early as the eighteenth century, attempts were made to develop guns to increase the darting power and range of harpoons; however, American sperm whalers found the early prototypes to be less effective than the primitive methods and only adopted advanced designs in the 1860s (Davis et al. 1987:59; Kugler 1971:21; Spears 1910:206). Instead, numerous variations were made to harpoon heads to improve effectiveness. During the early to mid-nineteenth century, harpoon tips used by pelagic whalers followed three main designs: the two-flued, the one-flued, and the toggle (Davis et al. 1987:60).

Throughout the early decades of the nineteenth century, the most common design was the two-flued iron. Sometimes called "primitive harpoons," this head style was sagittate, or arrowhead shaped, with sharp leading edges to cut into the whale and dull following edges to securely hook into the flesh and lodge there (Brown 1887:250; Davis et al. 1987:60; Giambarba 1967:53). Though whalers achieved a great deal of success with these, the design was not without its problems; many were unsatisfied with the tendency of two-flued irons to "draw," or work their way out, and let whales escape (Brown 1887:250; Kaplan 1953:84; Kugler 1971:21; Spears 1910:207; Spence 1980:49). The desire to reduce head size and in turn the cutting area of two-flued tips led to the introduction of the one-flued design in the early 1820s. Essentially a one-armed version of the older pattern, these became popular around 1840 (Lytle 1984). Though the one-flued iron was considered superior to its predecessor, it was only favored for a short time due to the introduction of the toggle iron in 1848 (Davis et al. 1987:60). Sometimes referred to as Temple irons after the African American whalecraft maker who first produced them, toggle heads were inspired by harpoons tips used by Indigenous hunters and incorporated a pointed barb sharpened on one side that was attached to the top of the shank with a pin at its center (Brown 1883:7; Harwood 1935:158; Kugler 1971:21; Spence 1980:114). When used, the barb entered the whale parallel with the shank; upon pulling back, however, it pivoted at a right angle to the shank, and the head became transfixed in the flesh (Brown 1883:7; Harwood 1935:158). The effectiveness of the toggle iron was quickly realized and soon after universally accepted among American whalers (Harwood 1935:158).

Since attempts at capture were sometimes unsuccessful, encounters occasionally resulted in whales escaping with harpoons still attached. To ensure that opportunities were fully attended to, it was standard for each whaleboat to carry four to five harpoons every time it was lowered (Olmsted 1841:20). Though two or three of these were intended as spares, the other two were referred to as "live irons" and attached to the whale line for immediate use when in pursuit (Leavitt 1973:18; Olmsted 1841:20; Spears 1910:210). To distinguish ownership of a harpoon, it was custom to mark the head of each with initials designating the ship to which it belonged (Starbuck 1878:154). These marks were generally placed on the flat of the shank near the head and made either by stamping or chiseling when the iron was red-hot or by inscribing or etching at a later point (Lytle 1984; Macy 1877:262–263). Stamps generally indicated the name of the blacksmith that crafted them, while inscriptions or etchings most often signified the name of the ship, its port, and often the whaleboat to which it was assigned (Brown 1887:251–252; Edwards and Rattray 1932:62; Hellman 2011; Lytle 1984; Spears 1910:208–209; Starbuck 1878:154). If a whale was killed and a marked harpoon from a previous attempt to take it was found in the carcass, the whaler's rule of the chase stated that "marked craft claims the 'fish' so long as it is in the water, dead or alive" (Brown 1887:252; Macy 1877:262–263; Spears 1910:209).

Next in importance to the harpoon was the hand lance (Brown 1887:252). Also known as the "killing iron," the lance was a spear-like implement that was used by the whaleboat's mate to kill a harpooned whale (Davis et al. 1987:58–61; Francis 1991:50; Harwood 1935:160; Lytle 1984). As with the hand harpoon, lances were primitive tools dating to the earliest period of whaling, and their design changed little over time (Brown 1883:12). Its components included a barbless, flat, oval-shaped head sharpened on both sides and welded at its aft end to an iron shank. The shanks used for lances were approximately 1 m longer than those used for harpoons but also tapered out to form a socket at the lower end (Olmsted 1841:21). As a precision instrument for cutting into and out of flesh, the lance was ground to a razor edge from shank to point on each side. To achieve such a fine edge, they were manufactured from the best quality materials, which, until the introduction of steel just after the mid-1840s, was typically tough wrought iron (Ashley 1926:87; Brown 1883:12; Davis et al. 1987:61–62; Doane 1987:67). As with the harpoon, this tool was completed when a 1.82-m-long timber pole was affixed into the socket and a 14.5-m-long rope known as a "warp" was secured for retrieval after each jab (Ashley 1926:87; Brown 1883:12; Olmsted 1841:21; Spears 1910:211).

Despite the invention of various designs of explosive bomb lances, American whalers preferred the hand lance for killing whales until the latter part of the nineteenth century (Davis et al. 1987:62–63; Kugler 1971:21). Each time whaleboats were lowered, every one of them was equipped with three to four hand lances (Brown 1883:13, 1884:252; Leavitt 1973:18). While one of those was used as an active tool and readied as soon as a whale was harpooned, the other two or three were kept onboard should they be needed as spares. Since they were not left in the carcass it was unnecessary to distinguish the name of the ship or boat with inscriptions or etchings; lances were, however, generally stamped with the blacksmith's initials or symbol (Hellman 2011; Lytle 1984).

The desirable properties of the shanks used for both harpoons and lances were toughness and flexibility. These qualities were extremely important both when a whale was harpooned, since it ran at great speeds while turning and sounding to break free, and when it was lanced, since the death flurry caused them to twist and turn wildly in all directions. These intense actions often resulted in shanks being broken, twisted into fantastic shapes, and even reduced in diameter by "tractile force" (Harwood 1935:158–159; Spears 1910:208). As such they were manufactured from the highest quality wrought iron available to allow them to bend under pressure (Brown 1887:252; Davis et al. 1987:61).

Hickory saplings with the bark intact were the preferred timber for whalecraft poles on American ships due to their strength; the bark also provided a good handhold (Ashley 1926:86; Davis et al. 1987:59; Davis et al. 1997:288; Edwards and Rattray 1932:59–60; Spears 1910:206). Though new whalecraft were sometimes coated in orange or red paint to prevent corrosion, aside from small amounts that adhered to crevices or the blacksmith's stamps, this paint was usually cleaned off by sharpening (Lytle 1984). When not in use, the heads of both the harpoons and lances were covered with wooden sheaths to keep their edges from being dulled and to prevent accidental injury (Edwards and Rattray 1932:61). Harpoons were stowed forward in the whaleboat on top of the thwarts below the port side wales; lances lay in racks under the wales on the forward portion of the starboard side with latches or pins to keep them in place (Ansel 1983:65). Because harpoons and lances were often misshapen while hunting, American whaleboats were equipped with a hammer and a grooved port chock timber to allow them to be straightened as needed (Pearson 1983:46).

Approaching the whale quietly and positioning the boat was crucial; once it was forward of the whale's tail and aft of its field of vision (Figure 6.4), the mate instructed the boatsteerer to stand up and "give it to him" (Allen 1973:172; Hohman 1928:159; Weiss et al. 1974:84). As soon as the boat was aside the whale, the boatsteerer "darted it" with the first harpoon, and if time permitted, then a second one attached to the same line was planted in case the first pulled out (Brown 1887:259; Dodge 1971:7; Hegarty 1960:38; Hohman 1928:159). The harpoons were not intended to kill the whale but instead to fasten to it, and sometimes, several attempts—or even numerous boats—were necessary to secure it (Macy 1835:226). If the harpoon was successfully buried and attached solidly to the whale, the boatsteerer called out "all fast" or "clear to the hitches," and the mate quickly yelled "stern all" to signal the oarsmen to move the boat out of the range of the dangerous flying tail flukes (Allen 1973:172; Grant 1932:66; Leavitt 1973:21; Stackpole 1967:41; WPA 1938:22). Meanwhile the boatsteerer and mate switched places in the boat so the mate could prepare to kill the whale using the razor-sharp lance (Allen 1973:172; Hegarty 1960:39; Stackpole 1967:41).

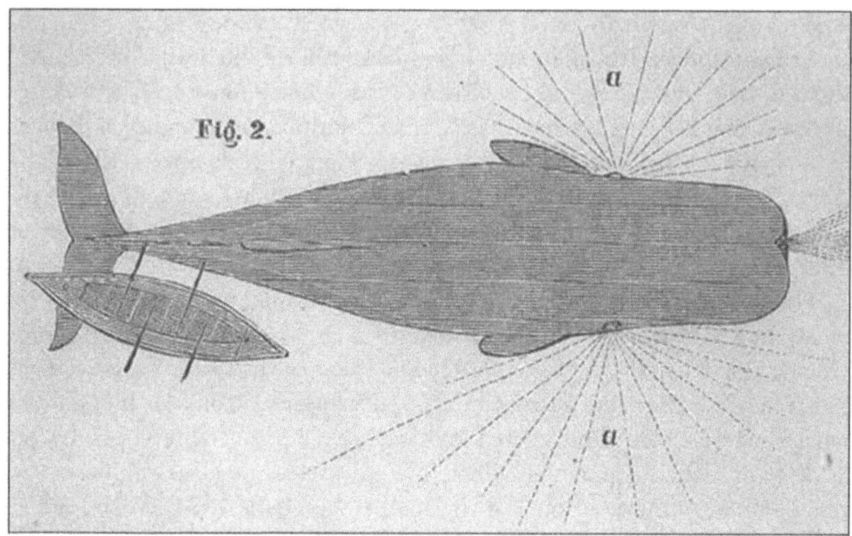

Figure 6.4. Top-down illustration of the approach for a whaleboat. The dotted lines indicate the whale's field of vision. By avoiding visual detection, the boat crew might avoid angering the whale until just before the harpoon was thrown. (From Davis 1874.)

Though not naturally aggressive, sperm whales were known to become violent when attacked (Martin 1974:8). Once harpooned, the whale might respond by running wildly, sounding, turning to attack the boat with its jaws, or breaching (Grant 1932:68; Stackpole 1967:43). If it ran, the whale line was taken out at a terrific speed, and the mate did his best to keep tension on the line and control the "Nantucket sleigh ride" that ensued (Hegarty 1960:39; Hohman 1928:159; Leavitt 1973:21; WPA 1938:22). All the while, one of the rowers used a small bucket to pour water over the rope while it ran over the loggerhead to keep the friction from causing it to smoke (Doane 1987:69; Verrill 1916:106). The literature on whaling is filled with instances of fouled lines causing boats to capsize, of broken or improperly rigged lines leading to crew members being maimed or killed, of boats being pulled deep under the water when whales sounded too quickly, and of whales crushing boats and crews in their jaws or smashing them with their massive tail flukes. Because of these dangers, sperm whales were said by whalers to be "dangerous at both ends" (for specific examples of the dangers, see Brown [1884:263–277] or Verrill [1916:115–162]).

Whatever the whale's reaction, when it eventually became exhausted the line was slowly hauled in until the boat was close enough for the mate to thrust a hand lance deep into its torso and "churned" up and down to find vital organs (Cheever 1850:53; Doane 1987:69–70; Dulles 1933:122; Leavitt 1973:23; Spears 1910:211; Weiss et al. 1974:84). Once the vitals were hit, the whale reacted with a sudden quivering of its body, emitted spasmodic spouts of blood—referred to by whalers as "flying the red flag"—and went into a final wild flurry before dying (Brown 1887:267; Doane 1987:70; Dulles 1933:124; Leavitt 1973:23; Stackpole 1967:43). Once dead, the whale was triumphantly referred to as being "fin out" or having "given up the ghost" and its body floated on the surface (Brown 1887:267; Dodge 1882:24; Shapiro 1959:35–36; Weiss et al. 1974:84; WPA 1938:23). If other ships were in the area and more whales were present, the dead sperm whale was "waifed" to deter theft while others were pursued; this marking was done by inserting a 1.5-m-long pole with a cloth flag into the carcass to indicate ownership (Delano 1846:81; Grant 1932:78; Hohman 1928:165–66; Olmsted 1841:58; Shapiro 1959:35).

Once a whale was killed, it was necessary to get the carcass back to the ship. If conditions permitted, the ship sailed as close as possible and took the whale alongside; if it was too rough or there was no wind, then it was necessary to tow the carcass to the ship (Doane 1987:70; Dulles 1933:125–126). Before transport began, the carcass had to be secured, which was done by first using a tool called a short-handled boat spade to cut a small hole either in the head near the blowhole or in the "small" at the root of the tail and then connecting a length of towline called a fluke strap (Brown 1887:271–272; Doane 1987:70; Grant 1932:78; Hawes 1924:168; Nordhoff 1874:121; Shapiro 1959:36; Stackpole 1967:43; Verrill 1916:111). To the fluke strap, a weighted line was attached, which was then connected to the boat via a towline secured to the loggerhead (Stackpole 1967:43; Verrill 1916:111).

Depending on the distance and the size of the whale, towing was an arduous task and sometimes took many hours (Hohman 1928:166). When the boats arrived at the ship, the carcass was brought along the starboard side with the flukes toward the bow and head toward the stern (Dulles 1933:126; Weiss et al. 1974:84). The carcass was then secured by running a heavy chain around the "small," through the "fluke pipe" in the ship's bulwark, and over the "fluke bitt" on the foredeck (Cheever 1850:53; Doane 1987:63; Leavitt 1973:124; Stackpole 1967:31). With the carcass securely attached, the sails were reefed to halt the ship and all boats were hoisted into their cradles (Delano 1846:26). The captain then ordered the cook to "supper the watch" or "dinner the watch." Once the meal was finished, the crew immediately began processing the whale, or if too late in the day, they prepared the ship to begin processing operations first thing the following morning (Brown 1887:278; Nordhoff 1874:122).

Archaeological Evidence of Raising Whales and Capture

Numerous pieces of whalecraft were identified among the remains of *Two Brothers* at French Frigate Shoals. These implements included harpoon heads, lance heads, and sections of whalecraft shanks. It is highly possible that other pieces of whalecraft are present at the site but are hidden due to thick layers of marine growth and iron concretion. All these artifacts were found scattered in Section B of the shipwreck site, indicating an association with the stern of the vessel. It is, however, impossible to determine whether they were spare implements that were stowed away or active pieces kept in the racks mounted directly under the skids or in boats that did not survive the wrecking. Undoubtedly, the tendency for whalecraft to be lost or damaged meant that ships were equipped with many harpoons and lances for each cruise. For instance, the accounts book for the 1820 to 1824 cruise of the Nantucket ship *Peru* lists at least 85 irons and 36 lances (NHA 1820). The outfitting book for the 1832 cruise of the New Bedford ship *Condor* is even more specific; for whale irons it lists "100 two flue 30 one" and between 25 and 30 lances (NBWM 1832). Notes found in the back of the log from *Parker*'s 1831–1835 cruise indicate that 150 harpoons and 30 lances were shipped onboard (Brown 1835:257).

Of the five harpoon heads located on-site, all are two-flued designs, which is consistent with wrecking in 1823. Each harpoon is missing the shank, and it appears that each separated at the spot at which it was welded together. The conditions of these artifacts varied; some are easily discernible by their sagittate form, while others are harder to identify due to corrosion and deterioration. The lance heads from *Two Brothers* each shared the standard elongated, oval-shaped design typical of hand lances. The condition of these artifacts is relatively good, and two of them include a small section of shank attached to their lower ends, while the other is missing the shank altogether. At least five sections of whalecraft shank were also found at the site. For the most part, the poor condition of these short, broken sections prevents a determination of whether they were associated with harpoons or lances. One specimen, however, appears to be largely intact and matches the length specified for harpoon shanks (Spears 1910:206).

Because the harpoon and lance heads are encased in calcium carbonate marine deposits, it was thought possible that any markings made on their surfaces could remain. This diagnostic potential coupled with a plan to interpret and publicly display them resulted in recovery of three harpoon heads and two lance heads. Each of these artifacts was determined to be composed of iron before being analyzed using CT-scanning technology to assess structural integrity and any intentionally made surface marking (Figure 6.5). Each artifact suffered considerable corrosion from long-term exposure to the marine environment, and all were assessed as being in fair to poor condition with no markings on their surfaces (Fox 2012:3). All recovered specimens then underwent full conservation treatment for eventual display.

Figure 6.5. Postconservation photograph of two of the three harpoon heads recovered from the *Two Brothers* site for interpretation purposes. (Courtesy of Papahanaumokuakea Marine National Monument.)

A concreted iron artifact identified in the sandy channel between the two sections of the *Two Brothers* shipwreck site could be the top section of a masthead waif. Though encased in corrosion, the shape of the object is easily discernible—a complete circle (20 cm in diameter) made of relatively thin metal strapping with a metal stem projecting from one side. Although the lower portion of the stem is deteriorated and appears to be incomplete, this object is largely intact.

Cutting In

Processing the whale was done as quickly as possible and carried out continuously until complete (Weiss et al. 1974:84). The first step in processing was known as "cutting in." Historian John F. Leavitt (1973:123) defined this as "the process of cutting the blubber or fat away from the carcass of the whale as it lies in the water while it was alongside the ship." Due to the large size of sperm whales, this operation was a formidable undertaking and involved specific tools to precisely procure the blubber with as little waste as possible. Prior to beginning this process, the cutting and lifting equipment was set up to ensure that it went as seamlessly as possible. Thus, when the shipkeeper determined that a whale had been taken, the ship erupted in a flurry of activity. Under favorable weather conditions and with skilled workers employed, an average sperm whale took approximately six to eight hours to prepare; however, in rough weather, this same process could take as long five days (Brown 1887:283).

The steps taken during the fitting out period, when the masts and working area on the ship's starboard side set up, greatly increased the efficiency of cutting in. First, the heavy cutting tackles and immense blubber hooks stored below decks were sent aloft to be attached to the previously rigged necklace or pendant on the mainmast and guy tackles on the foremasts, and then the 10 to 15 cm in diameter manila rope called the "falls" was threaded through them and run ahead to the windlass (Brown 1887:277; Davis 1874:77; Hazen 1854:84; Leavitt 1973:25; Perkins 1854:61; Scammon 1874:232). While this was happening, the removable section of the bulwark was taken out, the hatch to the blubber room removed, and the carpenter erected the cutting stage that was then hung over the side (Doane 1987:70; Perkins 1854:61). The cutting utensils were also retrieved and resharpened on the grindstone; these included cutting spades, boarding knives, pikes, and gaff hooks, all of which were made of iron and fixed to oak or hickory poles of varying lengths (Browne 1850:57; Olmsted 1841:62; Perkins 1854:61). Once the tools were sharp and the stations were prepared, the crew took their places and cutting operations began.

During the cutting in process, each member of the crew was tasked with specific jobs based on rank (Dulles 1933:126). Precision and experience were required, and the officers and boatsteerers took roles of responsibility: "The captain, first and second mates worked on the cutting stage; the third mate, stationed in the waist, was in charge of hoisting and stowing away the blanket-pieces; the fourth mate divided his attention between the waist and the windlass; and the boatsteerers performed the most exacting tasks in the blubber room, in the waist, and on the whale" (Hohman 1928:171). The foremast hands performed several duties including heaving on the windlass, helping in the blubber room, handling a "scoop net" to catch scraps of blubber that fell in the water, and guarding the boatsteerer when on the whale. One crew member was always detailed to turn the grindstone since the implements constantly needed resharpening (Hohman 1928:171).

The methods and patterns used for cutting in sperm whales were perfected by American whalers in the later decades of the eighteenth century, and illustrations depicting this (Figure 6.6) appeared by the 1790s (Ashley 1926:96; Dyer 1999). The primary task of cutting in was removing the large head and securing it to the side of the ship. A sperm whale's head constitutes approximately one-third of its length and contains a large cavity filled with the highest-quality oil, known as spermaceti (Brown 1887:278; Davis 1874:82; Stackpole 1967:50). The captain and mates accomplished decapitating the animal by using cutting spades to trace the incision pattern that followed the edge of the skull (just behind the eyes) and then cutting a hole to attach a blubber hook. The crew then wound the windlass to create strain and roll the carcass over to complete the cut and expose the joint between the skull and the vertebrae; the strain caused by the windlass and the spades chopping worked together to separate this joint. Chain strops were then roved through the cuts in the head and through a hawsepipe aft of the gangway to hang it securely over the ship's side while the rest of the carcass was processed (Davis 1874:82; Leavitt 1973:25; Olmsted 1841:64).

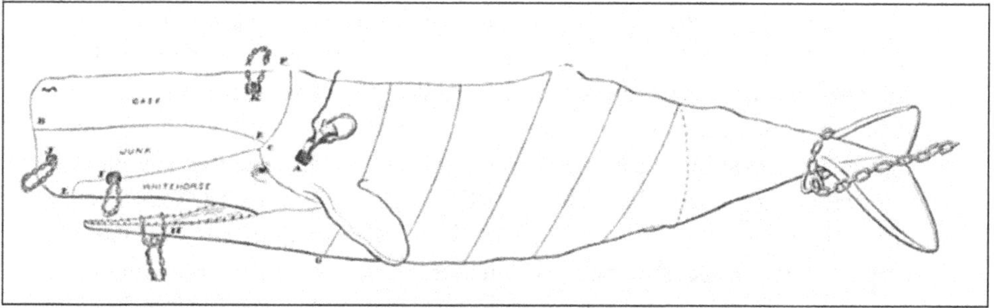

Figure 6.6. Illustration of the spiral cutting pattern used for removing blubber from sperm whales. (From Scammon 1874.)

As the head was removed, the second mate began "scarphing" the body; this involved cutting incisions diagonally into the blubber (Browne 1850:60; Doane 1987:71; Hohman 1928:167; Olmsted 1841:64). A hole was then cut into the blubber between the left eye and fin using a boarding knife, and a tethered boatsteerer went overboard to guide the blubber hook into it. With the blubber hook securely in place, six or eight of the crew cranked the windlass, which in turn rolled the carcass, and, with the help of the mates and their spades, stripped the blubber away in the same way that an apple or orange might be peeled (Delano 1846:26; Dulles 1933:127; Leavitt 1973:25; Shapiro 1959:37; Stackpole 1967:49; Weiss et al. 1974:86; WPA 1938:23). Once the resulting strip of blubber reached the upper blocks, a boatsteerer cut a hole at the lower end and inserted into it a second blubber hook or a wooden toggle attached to the second set of cutting gear. Once both were under tension, the upper portion of blubber was severed above the second hook using the boarding knife, and the free section—known as a "blanket piece" and weighing around a ton—was dropped through the hatch to the blubber room floor between decks. In the blubber room, two crew members had the formidable task of reducing the blanket pieces into smaller sections called "horse pieces," which they stowed in preparation for trying out. This process continued until the blubber was completely stripped from the carcass (Brown 1887:278–279; Browne 1850:128; Cheever 1850:58; Doane 1987:71; Dulles 1933:127; Leavitt 1973:25; Macy 1835:227; Olmsted 1841:64; Payne n.d.:4–5; Scammon 1874:236; Shapiro 1959:37; Stackpole 1967:49; Weiss et al. 1974:86). After most of the blubber was removed, a mate searched the inside of the stomach for ambergris, an extremely rare and pungent substance used as a base for perfumes and sold for the same price as gold (Hegarty 1960:40; Shapiro 1959:38).

Once the carcass was stripped and clear of ambergris, the tail was cut through and the remains set adrift to be devoured by the many sharks and flocks of screaming birds that had surrounded the scene from the start (Brown 1887:279; Delano 1846:27; Dulles 1933:127; Leavitt 1973:25; Payne n.d.:5; Scammon 1874:235). Attention then returned to the head, which was hooked and brought forward to the gangway, hoisted nose down, and either set onto the deck for dissection or, if it was too large, suspended over the side out of the water as high as possible (Browne 1850:128–129; Leavitt 1973:25; Payne n.d.:5). Three parts of the sperm whale's head were of value to the whalers: the upper part termed the "case," which was the largest cavity and contained the purest oil (spermaceti); the lower half of the forehead called the "junk," which was filled with oil and spermaceti; and the lower jaw with its numerous teeth, which were prized by whalers for use in scrimshaw and for trading with Pacific Islanders (Davis 1874:82; Delano 1846:27; Hohman 1928:169; Payne n.d.:5–6; Shapiro 1959:36).

Processing the head started with separating the jaw, which was set aside for later removal of the teeth using the cutting tackle (Davis et al. 1997:276; Hohman 1928:169; Weiss et al. 1974:87). The junk was then cut from the skull and secured to the deck to ensure that it did not slide as the ship rolled. Next, the case was breached by cutting a hole the size of a barrel through a thick membrane at the base of the skull known as the "white horse." Once opened, the spermaceti was bailed from the case. Bailing was achieved either by using a "tail block" with a line and "case bucket" rigged to the cutting tackle or by a crew member entering and passing case buckets out (Delano 1846:27; Grant 1932:96; Olmsted 1841:65; Leavitt 1973:27; Stackpole 1967:49–50). Generally, between 15 and 30 barrels of spermaceti were recovered from the case, which in warm weather is a liquid but in cold has the consistency of lard. It was then carefully stored in special casks and kept separate from the rest of the oil (Dodge 1882:9; Dulles 1933:128; Grant 1932:96; Stackpole 1967:50).

Once the case was emptied and the head pushed overboard, the junk was processed; this clear, fatty, spongy, oily substance was as rich in spermaceti as the case and ranked next in price (Doane 1987:73; Hohman 1928:169). It was cut into varying sized pieces, the larger having the oil removed by boiling and the smaller being squeezed out by hand (Doane 1987:73).

Archaeological Evidence of Cutting In

A small number of artifacts associated with the cutting in process were identified at the *Two Brothers* and *Parker* shipwreck sites. As the preceding discussion illustrates, the tools needed to carry out this stage of pelagic whaling included highly specialized implements such as blubber hooks and cutting tackle, while others like the previously discussed ship's windlass were multipurpose objects that might be found on any wrecked sailing ship of the period. Thus, the location of associated objects and their interpretation as a group provides insight into their use onboard a working whaleship of the early nineteenth century.

One artifact found on the *Two Brothers* site and identified specifically as cutting in equipment was a deteriorated block. Blocks are essentially pulleys that assist in lifting objects by increasing purchase (Biddlecombe 1925:3). These simple wooden devices were composed of a heavy outer shell, at least one inner wheel known as a sheave that was grooved to seat a rope, and a pin that acted as an axle on which the sheave turned. The sides of a block's outer shells were generally grooved at the top and bottom to receive a strap called a "strop," which was used both to reinforce the shell and to provide attachment points (Stone 1993:74–75). Since all blocks used onboard sailing ships had to be extremely tough to cope with excessive strain and the harsh conditions experienced, they were generally made from hard, heavy woods such as lignum vitae, ash, or hickory (Stone 1993:74). Though many blocks of varying sizes and purposes were found on all early nineteenth-century sailing ships, those used for cutting tackle were easily distinguished by size. As described previously, the cutting tackle consisted of a single, large double-block known as the "lower block," which was used for primary lifting; two large, single-blocks called the "upper blocks," which were used for the secondary lifting; and a slightly smaller "guy block" through which the guy line was rove. Historic descriptions of these blocks are rare; for the most part whaling historians have drawn from Captain Charles M. Scammon's extensive discussion of whaling gear (Scammon 1874:216–240), but he provided no detailed information about cutting tackle. Some useful information about blocks is, however, provided in James Temple Brown's description of those included in a display about the American whale fishery installed at the 1883 International Fishery Exhibition (Brown 1883). In a booklet intended as a companion to the exhibit, Brown indicates that although rope was the earliest means for strapping lower blocks, by at least 1883, it was replaced by "the improved chain strop" (Brown 1883:63). He also provides the dimensions for the cutting tackle from the display as, "lower block 18 by 12 by 10 inches [46 × 30 × 25 cm]; upper blocks 18 by 12 by 6 inches [46 × 30 × 15 cm]; guy block 13 by 9 by 6 inches [33 × 23 × 15 cm]" (Brown 1883:63).

The block identified at the *Two Brothers* shipwreck site is partially intact due to the hardwood from which it was manufactured. Though heavily deteriorated, the remains include two pieces of the block's outer shell and a single wooden sheave, which are attached to the pin. Since most of it is missing, overall length and thickness dimensions could not be ascertained; however, the intact pin allowed for the artifact's width to be determined as approximately 30 cm. Based on this measurement, the lack of any evidence of chain adhering to it, and the presence of only one sheave, these remains were determined to belong to one of the two upper blocks of a cutting tackle.

The other artifact specifically designed for and used in the cutting in process was the blubber hook. Among the most iconic implements associated with the historic whale fishery, these large iron hooks were used for securing blanket pieces while the blubber was stripped from the whale's carcass. Produced in blacksmith shops using high-quality iron, they measured roughly 76 cm in height and weighed between 31 to 68 kg depending on the size of the ship (Brown 1883:20; Olmsted 1841:62). Though variation in design occurred, they were basically J-shaped objects with an eye at their upper end for attaching to the lower block using a shackle or thick rope. A loose ring generally affixed into a small hole in the heel of the hook was used to attach a thin rope that allowed the point to be directed into the hole cut in the blubber for lifting (Brown 1883:64). In practice, blubber hooks were interchangeable with large wooden toggles often used for the same purpose. Hooks, however, were chosen for primary lifting during shipboard cutting in operations. Despite being reinforced with extra-thick iron in the throat (the curve at the bottom of the hook), blubber hooks occasionally broke under the immense strain of lifting blanket pieces (Brown 1883:64). Thus, it was standard for ships to carry several spares onboard. A review of available whaleship outfitting data for the period from 1800 to 1840 suggests that ships went to sea equipped with three to four blubber hooks listed among their outfits (NBFPL 1840; NBWM 1800, 1832; NHA 1820).

Three large blubber hooks were identified among the artifact assemblage of *Two Brothers*. Two of these have almost matching dimensions and are identical in design, with features including a reinforced throat and an upper eye that was forged in the manufacturing process. While the other hook has similar dimensions and includes reinforcement in the throat, the eye is made of a separate iron ring that was inserted into a hole in the top of the hook's body and would have allowed it to pivot. Marine encrustation on each of these hooks prevented identification of any markings or holes in their heels. All three blubber hooks were found in Section B of the site with two of them lying adjacent in a deep pocket at the edge of the reef (Figure 6.7) and the other in a small gulley on top of the reef. The proximity of the two suggests they were spares kept in storage in the stern of the vessel, while the location of the other hook could indicate that it was in use at the time of loss.

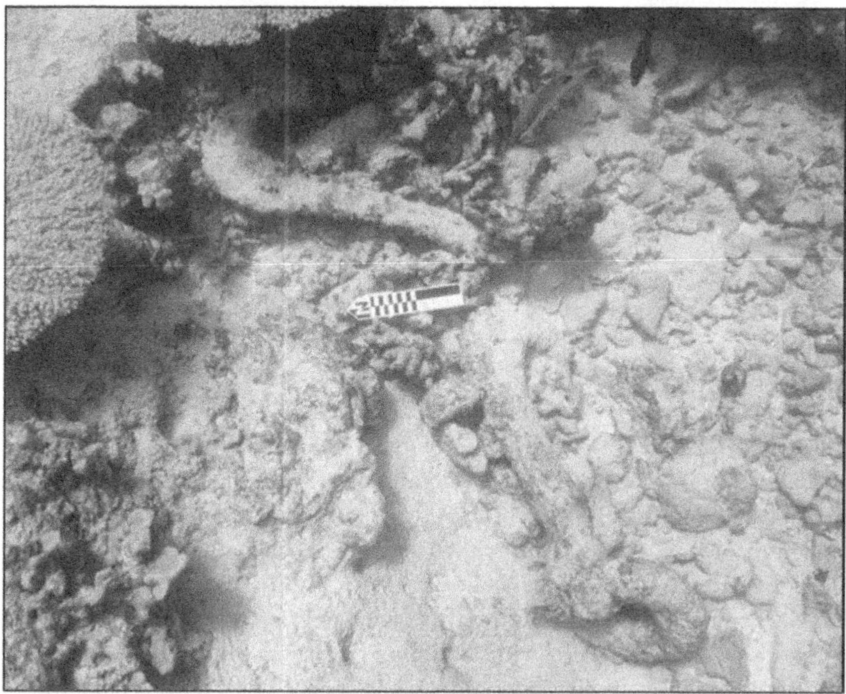

Figure 6.7. Two blubber hooks recorded in a pocket in the reef at Section B of the *Two Brothers* shipwreck site. (Courtesy of Papahanaumokuakea Marine National Monument.)

A single blubber hook was recorded at the *Parker* shipwreck site. This hook is basically J-shaped and approximately 61 cm in height. Though similar in design to the one found on *Two Brothers* with the separate ring affixed to its top, this hook lacked the same reinforcement in the throat and instead employed a heavy shackle with the pin inserted into the hole at the top of the hook's body. Marine encrustation and iron concretion on the back side of this hook prevented identification of possible markings or hole in its heel. Notes pertaining to outfits found in the log from *Parker*'s 1831–1835 cruises list four blubber hooks as being onboard (Brown 1835:257). The fact that this artifact was found in the bow section of the wreck near the remains of the windlass may indicate that it was an actively used implement, temporarily stored on deck at the time the ship came to grief.

A subsidiary component of the cutting gear was the iron thimble or "grommet" (Brown 1883:63; Higgins 1927:35). As discussed in the preceding chapter, these rings were added to eyes made in rigging lines to keep their openings from cinching and to prevent chafing. Although varying sized thimbles were found throughout the rigging of all working whaleships of the period, their ability to prevent distortion of the eye when under heavy strain meant that they were particularly useful for the cutting gear. An illustration of cutting in tackle in A. Hyatt Verrill's *The Real Story of the Whaler* (1916) indicates that thimbles were added to at least three specific sections: at the lower end of the manila strapping that surrounded the lower block, in the eye for attaching the blubber hook; above the upper blocks, where a smaller one was inserted for connecting them to the pendant shackles (Figure 6.8); and at the end of the small line shackled to the ring of the back of the blubber hook (Kipping 1859:86–87; Verrill 1916:41). Thimbles were also added at the ends of the thick cables that made up the pendants and from which the cutting tackle hung (Brown 1887:281).

Several thimbles were documented at both the *Two Brothers* and *Parker* shipwreck sites. Because they were ubiquitous objects, the small diameters of those artifacts generally prevented determining where specifically they were used on the ships. A particularly large thimble documented at each of the sites was probably incorporated into either the cutting tackles or the pendant. The large diameters and width of the discernible grooves in the outer faces of these heavily concreted artifacts indicate that each would have been used for seating a very thick rope that could withstand immense strain without breaking. Both are considered to have been part of the cutting gear, but their positions within them are uncertain. The thimble documented at *Two Brothers* was found in Section A of the site among the dense scatter of iron rigging elements and could have been inserted at one end of the pendants, which remained around the mast until the last whale was taken (Brown 1887:281). Since the one found at *Parker* was resting among the artifacts associated with ship's bow and near the blubber hook, it is suspected to be part of a cutting tackle.

Other objects recorded at the shipwreck sites were also directly used for cutting in. These include the remains of a windlass on the *Parker* wreck and a grindstone from *Two Brothers*. Though each was discussed previously, it is important to note that onboard whaleships these objects performed multiple functions. For instance, it is well-known that the windlass was a critical part of all nineteenth-century ship anchoring systems. While whaleships were

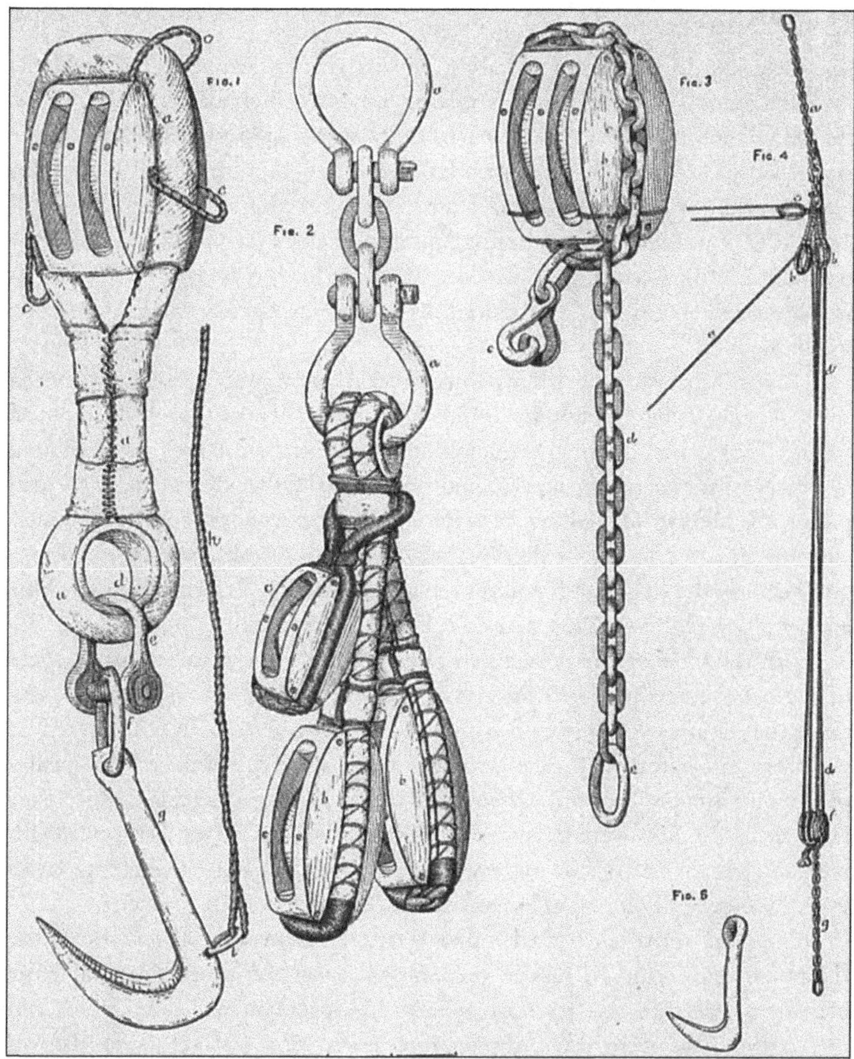

Figure 6.8. Historical illustration of the various components of cutting tackle including blubber hooks, upper and lower blocks, and thimbles. (From Verrill 1916.)

equipped with one for that purpose, it also functioned as a critical component of the cutting in equipment and without its adaptation for that purpose, blubber simply could not have been brought onboard. As for grindstones, while all wooden sailing vessels required them to keep blades sharp, the array of specialized implements needed for pelagic whaling resulted in their almost constant spinning—a task most lamentable by foremast hands (Browne 1850:131–132).

Trying Out

The next step in the pelagic whaling process was known as trying out. This involved removing water and extracting oil from blubber by boiling it in the large cauldrons situated in the brick tryworks (Doane 1987:266; Olmsted 1841:67). Trying out a large whale took generally three full days, with crews working six-hour watches—known as the trying out watch (Nordhoff 1874:128)—and operations running around the clock (Olmsted 1841:67; Robotti 1962:113). According to whaling historian John F. Leavitt (1973:28), an average-sized sperm whale produced 40 to 50 barrels of oil and a large one might yield 80.

As with other steps in the whaling process, crews were closely supervised since mistakes could result in diminished oil quality and, in turn, reduced profits. Thus, when the pots were boiling, a mate or boatsteerer was relegated as officer of the watch to superintend operations. Other duties included preparing the blubber for boiling, carrying it from the blubber room to the deck, cleaning the deck to reduce slipping hazards, moving casks, maintaining lookout, keeping the helm, and lending a hand as needed (Brown 1887:285; Doane 1987:73; Hohman 1928:172–173; Payne n.d.:6).

With the blubber and spermaceti onboard and the carcass cut loose, some of the boatsteerers prepared for trying out and readied the implements. Because the try-pots were covered with wooden lids when not in use, it was first necessary to clean their insides with soapstone since rust reduced the quality of the oil (Browne 1850:60; Cheever 1850:87; Palmer 1959:145). Once they were clean, the fires were started using a small amount of wood or coal. With preparations for trying out underway, the cooper was busy assembling casks from shooks, head pieces, and metal hoops (Grant 1932:40).

While the tryworks heated, other boatsteerers oversaw the work of the "blubber room gang." Blubber preparation consisted of cutting the large blanket pieces into smaller horse pieces 15–46 cm in width by 30–61 cm in length; these were then thrown into open tubs and set aside (Brown 1887:285; Doane 1987:73; Dulles 1933:129; Payne n.d.:6; WPA 1938:23). As needed, the horse pieces were brought onto the deck and taken to a wooden table with a rounded top known as the "mincing horse" or "mincing block." There they were sliced using a two-handled blade called a mincing knife to provide more surface area for quicker boiling (Allen 1973:178; Cheever 1850:61; Macy 1835:228; Verrill 1916:49; Weiss et al. 1974:87). Care was taken to ensure that the slices were thin but did not quite go all the way through the horse piece; when done correctly, the minced pieces resembled the pages of a book and were referred to as "bible leaves" (Doane 1987:73; Grant 1932:92; Payne n.d.:6; Robotti 1962:112; Shapiro 1959:39;

WPA 1938:23). Once prepared, the bible leaves were thrown into a mincing tub near the tryworks to await boiling.

As the blubber boiled, any pieces of skin floating on the oil were skimmed off using a perforated tool called a strainer and added to the fire as fuel. These brown "doughnut-looking" pieces called scraps or cracklings were saturated with oil. As such, they produced an intense heat and black smoke that stained the sails (Browne 1850:56; Davis 1874:89; Davis et al. 1997:274; Doane 1987:73; Dulles 1933:129; Hohman 1928:172–173; Payne n.d.:6; Shapiro 1959:39; Verrill 1916:50; Weiss et al. 1974:87). At night, scraps were placed in an iron basket called a cresset and burned for light (Allen 1973:178; WPA 1938:24). Some of the scraps were set aside in a "scrap hopper" to be used later for fires, while the rest were either thrown overboard or saved to be traded with South Sea Islanders who enjoyed eating them, as did some of the whalers (Doane 1987:73; Olmsted 1841:67; Scammon 1874:238).

If the spermaceti from the head was to be tried out while onboard, it was cooked together with the junk prior to the blubber being added to the pots. Segregating the raw materials prevented mixing it with lower quality oil, which guaranteed the best price. When complete, this oil was placed into special casks, marked as "head-matter" and stowed separately from the rest of the oil (Scammon 1874:239; Weiss et al. 1974:87). Once trying the junk and spermaceti was done, the bible leaves in the mincing tubs were fed into the try-pots using a two-pronged blubber fork (Browne 1850:61). Approximately one hour was required for a pot full of blubber to be rendered, with frequent stirring to ensure that it did not burn or settle in the bottom of the try-pots (Doane 1987:73; Robotti 1962:112; Scammon 1874:238).

When the oil had separated from the water and flesh, a long-handled iron or copper bailer was used to scoop it from the pot and pour it into a rectangular copper tank situated at the side of the tryworks (Browne 1850:56; Davis et al. 1997:274; Grant 1932:94; Hohman 1928:172; Macy 1835:228; Scammon 1874:238; Stackpole 1967:50–51). Called a cooler, this tank provided a valuable function since boiling oil poured directly into a wooden cask would have caused it to shrink and leak (Doane 1987:73–74; Olmsted 1841:67). A cooler held 6 to 10 barrels of oil and included a screen-covered valve called a "stopcock" for transferring oil into storage containers (Olmsted 1841:67; Robotti 1962:112). After all the blubber was tried, the fires under the pots were extinguished, and attention shifted to storing and stowing the oil (Nordhoff 1874:131).

Archaeological Evidence of Trying Out

As discussed in the previous chapter, the inclusion of onboard tryworks made trying out perhaps the most iconic activity associated with pelagic whaling since the mere sight of a plume of black smoke rising from a ship on the horizon alerted passersby to its function. In much the same way, the presence of numerous bricks and/or large iron pots found among the remains of a shipwreck site are obvious indicators of a lost whaleship. Although many of the industrial artifacts from this period were made of materials that deteriorated quickly or were swept away after the initial wrecking, the dense clay bricks and heavy iron pots tend to settle on the seabed after the tryworks separate. As such, these components are part of the artifact assemblage of nearly every wreck of a pelagic whaleship so far identified, and their diagnostic value cannot be understated. Though other tryworks components such as iron knees might also be expected to remain, those are less often found on archaeological sites. Since tryworks knees were bolted to the deck, unlike the bricks and pots that scattered freely, the knees likely floated along with the wooden structure.

The most prevalent component of the tryworks documented at American whaling shipwreck sites are undoubtedly bricks. The only known historical information pertaining to tryworks bricks is found in an outfitting book dating to the 1860s. Included in the first pages, among data about various industrial equipment are specifications for the two types of bricks used onboard whaleships: "3 1/2 by 7 inches" (9 by 18 cm) and "4 inches by 8 inches" (10 cm by 20.5 cm) (Kirby 1860:5). As discussed, the tryworks were installed on whaleships during the precruise stage, so the number of bricks used in their construction went unnoted. The actual number, however, varied greatly depending on the dimensions and design of the structure. For a general sense of the actual numbers needed, 455 bricks were used to build the two-pot tryworks onboard the half model of the whaleship *Lagoda* at the New Bedford Whaling Museum in 1917 (Olly 2004:150).

Most ships carried spare tryworks bricks among their outfits, and some agents listed an exact figure in their outfitting books; for example, *Marcella* (1840) and *Mars* (1845) reported carrying 200 and 250 spare bricks respectively (NBFPL 1840, 1845). Others, like the whaleship *Condor*, appear to have been less concerned with the actual number and simply note "got a plenty" in the margin next to the entry "Spare Brick" (NBWM 1832). The implied need for carrying spare bricks and lime for mending a damaged tryworks is only one interpretation of their presence onboard. According to the mid-nineteenth-century narrative of whaler William B. Whitecar Jr., several bricks were kept in each of the whaleboats and used to heave at a whale to determine whether it was belligerent in nature and if engaging it should be avoided (Whitecar 1864:344).

The scattered remains of tryworks bricks were documented at both the *Two Brothers* and *Parker* shipwreck sites. No markings were noted on any of the bricks at either site, and measurements of intact samples from each were consistent with the specifications for the smaller of the two types described above (Kirby 1860:5). Since machine-made bricks did not appear until the 1860s (Smith et al. 2006), it is presumed that these were manufactured using wooden molds. The presence of brick at these sites confirms that they were both still engaged in fishing, since their tryworks had not yet been "knocked down" (Ashley 1926).

The other principal component of the tryworks were the try-pots. Originally known as try kettles, these large iron cauldrons of varying size were used to boil oil whale blubber. Although they were integral to operational success, detailed information pertaining to try-pots is rare. As with the tryworks bricks, the most useful data so far found comes from a blank outfitting book dating to the 1860s, which provides dimensions for two types: the "Old Pattern" and the "New Pattern." Although there is no indication of stylistic differences or of the period associated with the two types, the Old Pattern is shown to include four sizes measured in gallons (140, 160, 180, and 200) (530, 605, 680, and 760 liters), while the New Pattern includes two sizes measured in gallons (200 and 220). And while both types include a 200-gallon size, the noticeable differences in the length and width dimensions of the two could make distinguishing between them possible (Kirby 1860:5).

While most historic descriptions and illustrations of tryworks built on American whaleships indicate prevalence for two pots in the structure, three-pot tryworks were also used (Davis 1874:24; Macy 1835:228). The number of try-pots taken onboard a ship depended on its size, the size of the intended brick structure, and the pots available at the time of outfitting. It was common for each ship to carry a spare try-pot (Hegarty 1964:63), and many outfitting books state that two sets of try-pots were carried (NBFPL 1841; NBWM 1832). Thus, it would be typical to find as many as four pots onboard a whaleship of this period.

Along with the tryworks bricks, shards of broken try-pots were recorded at the *Parker* shipwreck site. Scattered along a trail leading through a small pass in the reef crest and throughout the back reef area are at least seven pieces of try-pots of varying size (Van Tilburg 2006). Although some large shards were noted among these, it was not possible to determine the liquid capacity of the pots. Their spatial distribution, and that of associated bricks, indicates that the tryworks were already breaking apart as the ship was pushed over the reef. Due to the almost constant, heavy swells that pound the reef at Kure Atoll, the try-pots were likely smashed to pieces in the depositional environment.

Four largely intact try-pots were documented at the wreck of *Two Brothers*. Scattered across both sections of the site, the try-pots rest in pockets in the reef that appears to have protected them from damaging swells. All four were assessed as being in fragile condition, however, the marine encrustation and iron concretion on each help with preservation. Although a portion of the rim and sidewall of two of the pots has collapsed, each retains its globular shape, as well as its flared rim, lifting eyes, pegged feet, and evidence of the decorative rings on the sides (Figure 6.9). Although concretion has somewhat distorted their original dimensions, data extrapolated from these try-pots indicate that their volumes range from 125 to 174 gallons, which closely aligns with the Old Pattern try-pots discussed above (Kirby 1860:5). Three try-pots were found scattered over the extent of Section A, which indicates that the ship was either equipped with a two-pot tryworks with a spare pot lashed to its side or that three pots were built into the brick structure. Based on Nantucket historian Macy's observation that tryworks on the island's early nineteenth-century whaleships were equipped with "two, and sometimes three pots" (Macy 1835:228), either of these configurations is plausible. The other try-pot, documented in Section B of the site, was most likely a spare since it was found far from the others with objects thought to have been stored in the stern of the ship.

Figure 6.9. Intact try-pot documented at Section B of the *Two Brothers* shipwreck site. (Courtesy of Papahanaumokuakea Marine National Monument.)

The final object documented at the *Two Brothers* site and thought to have been used for trying out is a hook found concreted to the reef in Section B. Initially interpreted as one of many generic hooks that would have been used at various points in a whaleship's rigging, the large size of this artifact suggests that it is instead a small blubber hook (Brown 1883:64). Also known as a "junk hook," this iron tool measures 23 cm in length and incorporates a small ring at its upper end for attaching a rope. According to James Temple Brown, these were used onboard whaleships prior to the 1880s for "handling blubber, clearing the hatch when it was blocked with blubber, and hauling the junk aft when was it is to be *lashed* [*sic*]; hence the name junk hook" (Brown 1883:64). Although no illustration of a junk hook accompanies Brown's description, the features of the artifact are the same, and it directly matches the size.

Stowing Down

After the oil cooled, it was necessary to shift it into casks for storage. As long as it was properly handled, whale oil could be stored for long periods without spoiling; thus, this task—often referred to as "stowing down" (Ashley 1926:98; Bronson 1855:52; Davis 1874:234; Delano 1846:75; Hohman 1928:175; Melville 1851:475–77; Nordhoff 1874:131; Stackpole 1967:51)—required the utmost care and attention to ensure that every drop of oil was preserved (Whipple 1979:41). A great deal of consideration was also given to the positioning of the casks in the hold. Since the object of the cruise was to return with as much oil as possible, it was necessary to make the most of the available space by carefully and precisely stowing casks. As such, no time was spared in stowing down, and often, it took several days to complete (Stackpole 1967:51). Few descriptions of the stowing down process are available; however, it appears that the two methods used in transferring oil into casks involved either a bailer or a hose (Leavitt 1973:27).

If more than one whale was captured, tryworks operations ran nonstop, and as a result, oil could only rest in the copper cooler for short time before a fresh batch was ready. Under such circumstances it was necessary to transfer the oil into wooden casks on deck to cool longer, which was accomplished by bailing (Hawes 1924:172). After being filled, these casks were lashed securely to the deck to reduce the chances of their sliding around and spilling or hurting someone as the ship rolled (Hohman 1928:173; Scammon 1874:238). Once the oil in them had cooled the casks shrank, and the hoops were driven farther down onto them to tighten their seams (Macy 1835:228; Olmsted 1841:67; Scammon 1874:238). In his reflections of whaling operations in the 1840s, Captain Benjamin Doane suggests that "no cask was considered fit to stow down until it had been three times to the tryworks, that is, until it had been filled with hot oil and set to cool three times" (1987:74). This comment supports statements made by other early to mid-nineteenth-century whalers regarding the extent of cask shrinkage and further illustrates the care that was taken to prevent the loss of oil through leakage (Macy 1835:228). Once properly coopered, a "gimlet hole" was made in each cask to allow it to vent; if this was not done, the casks had the potential to burst from expansion in warm climates or draw in their heads in cold weather (Doane 1987:74).

The other method for transferring oil from the cooler to casks involved the use of a hose. Referred to as "running down the oil" (Brown 1887:286), it was first necessary to let the liquid cool further by transferring it into a large deck pot sometimes referred to a cooler (Allen 1973:179; Leavitt 1973:27; Olmsted 1841:67; Scammon 1874:238). After the oil sufficiently cooled, it was necessary to then transfer it to the casks in which it would be stored for the remainder of the cruise. The easiest way to do this was attaching a leather or canvas hose to the copper or iron deck pot's stop valve (Allen 1973:179; Ashley 1926:98; Doane 1987:74) or by bailing oil into a funnel attached to a hose (Bullen 1898:40). The other end of the hose was then sent below where casks were filled and stowed (Allen 1973:179; Bullen 1898:40; Leavitt 1973:27; Stackpole 1967:51). As described earlier, a common practice onboard ships hunting sperm whales was to fill the largest casks with water and stow them on the lowest tier for use as ballast until they were needed for oil storage. Doane (1987:74) refers to this operation as "hosing down" and describes one way of doing it as follows: "When a considerable quantity of oil is accumulated and cooled, in barrels on deck or in the between deck, the cargo will be broken out until the ground tier casks are reached; the salt water is pumped out of such of them as it is intended to fill [*sic*] at the time, and they are swabbed out dry. A tub made by sawing one of these tun butts [*sic*] in half is placed on skids over the hatch, a hole is bored in the bottom of it, and the end of a canvas hose 50 to 75 feet [15–23 m] in length is tacked over the hole and the other end let down into the hold. The barrels of cool oil are then rolled up on the skids to the tubs, and the oil is allowed to run into it, whence it is directed through the hose into the ground-tier casks."

Regardless of which method was employed for filling casks, on vessels hunting sperm whales this process was overseen by the captain or first mate, who measured the hold to determine the best placement of casks and ordered the different sizes to be fetched from the cooper (Brown 1887:286). The first or second mate supervised the work on deck to make sure that it was completed as efficiently as possible; duties included overseeing one of the lesser officers or boatsteerers attending to filling, inspecting the cooperage, and directing the placement of casks on deck. The rest of the crew on the watch might be tasked with assisting in filling and/or moving casks and lashing them to the deck where they could cool longer, helping the cooper, or assisting in stowing the casks in the hold (Brown 1887:286). No matter the task, this work was filthy, and there was always a danger of slipping because, despite the best efforts to avoid it, oil inevitably found its way onto decks.

Once the oil in the full casks had sufficiently cooled, they were ready to be stowed. The process of stowage involved more than simply moving casks; it required preparation of the storage area and constant monitoring of progress to ensure that available space was maximized. Beginning with initial measurements of the lower hold, casks of varying sizes were sent down and positioned precisely before being chocked with wedges of wood to prevent movement. Moving these large and unwieldy containers was backbreaking and dangerous work, which required skill even in calm seas (Ashley 1926:98). Casks were hoisted using heavy lifting tackle and "can hooks" that gripped them at their chines, which was a much safer method than using slings (Ashley 1926:98). According to the outfitting book of the bark *Mars*, both small and large size can hooks were used on whaleships (NBFPL 1845:30), depending on the size of the cask to be lifted. Once secured, casks were passed down through the hatches and into the hold, where crew laid them their sides running fore and aft, then manipulated them into the assigned place.

Evidence of the actual sizes of the casks used and their positions is limited to two documents. The first of these, a one-page diagram from 1841 of one-quarter of the lower hold of the ship *Gratitude* (Figure 6.10), indicates that stowage was done in three tiers, with the second tier for most of the larger casks (Hussey n.d.). The second document, from an unspecified ship, is far more detailed and includes annotations. Although much of the text is illegible, it depicts the same stacking order for the lower hold through three cross section diagrams and illustrates the plans for the between decks area, which were used for stowage once the hold was filled (NBWM n.d.).

Archaeological Evidence of Stowing Down

Archaeological materials related directly to stowing down are difficult to identify at the *Two Brothers* and *Parker* wreck sites. Most components used for this purpose were of organic materials that would deteriorate rapidly after their deposition in the marine environment. Although outfitting records indicate that all ships of the period were equipped with hoses (leather or canvas) for transferring oil to the wooden casks, no physical evidence of such was found at either site. As for the metal implements used in this process, aside from an iron deck pot, most were small and likely became obscured in the reef environment once marine encrustation covered them. Although the can hooks used for lifting casks were an integral tool and a standard item on outfitting lists from at least 1820, none have been found archaeologically (NBFPL 1820).

Although the wooden components of the casks used for oil storage deteriorated quickly, remains of some of the metal hoops for binding them remain at both the *Two Brothers* and *Parker* sites. As aforementioned, thousands of hoops were shipped onboard whaling vessels for use with shooks and head pieces to construct casks as needed. Though the primary goal of whaling was

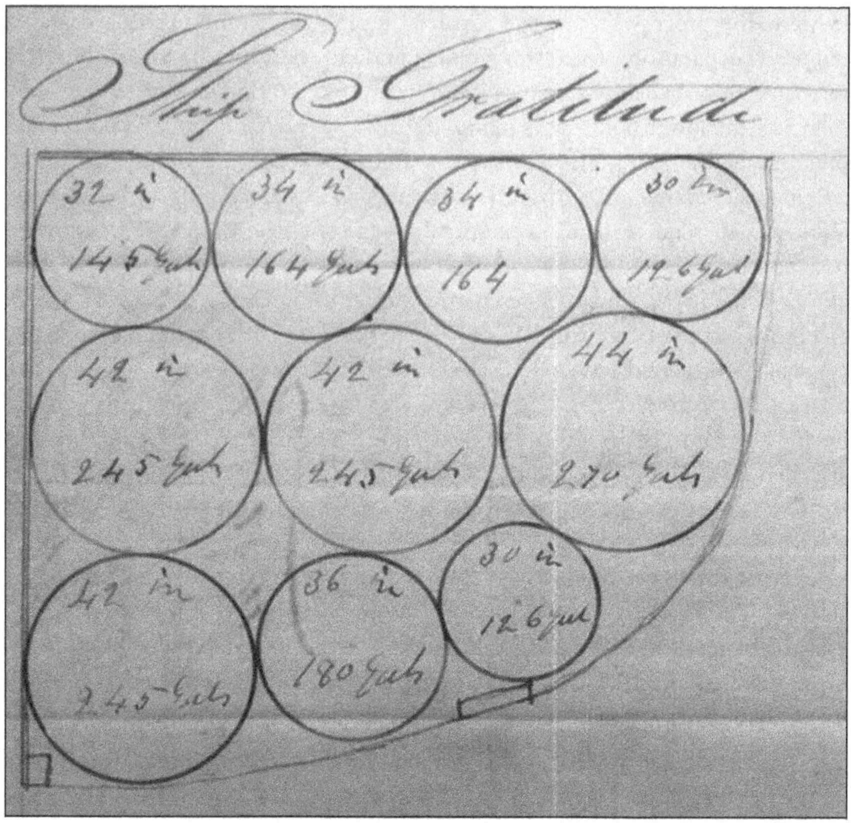

Figure 6.10. Diagram from 1841 showing the intended stowage of casks of varying diameters and volumes in a section of the whaleship *Gratitude*. (From Hussey n.d.)

to fill casks with oil, since so many were used to store food and other provisions it is impossible to determine exactly what was stored in the casks bound by them.

Artifacts found at the *Two Brothers* and *Parker* wreck sites related to lifting can also be associated with stowing down. To shift the immense weight of full oil casks, whalers employed the windlass, rigging blocks, thick manila ropes, and chains. Although these were by no means specific to this stage in the pelagic whaling process, they were nonetheless essential for ensuring that casks were not damaged while being stowed and that no oil leaked unnecessarily through mishandling. Since few detailed descriptions of stowing down are available, it is unknown exactly which lifting tools were used. It is unlikely, however, that the relatively small blocks found at the *Parker* site were used to lower heavy casks into the hold. Instead, it is probable that the large blocks of the cutting tackle, already rigged to the windlass for heavy lifting purposes, were utilized in stowing down since larger blocks and sheaves produced far greater power (Stone 1993:75).

Preparing the Ship for the Next Round of Whaling

Once the work of cutting in, trying out, and stowing down were complete, it was necessary to clean the ship and prepare all stations for the next time whales were raised. This final step in the pelagic whaling process was crucial since the lookouts were constantly manned and the potential for the next round of activity was ever present. Thus, the less prepared the ship, equipment, and crew, the greater the chance of missing a whale, which in turn meant a direct loss of profit and an extended period at sea.

The processing of carcasses onboard a whaleship created an absolute mess. Every exposed surface on and near the deck was coated in oil, blood, and offal (known as gurry), while smoke from the tryworks covered the masts and sails in oily soot (Grant 1932:98). Reverend Henry T. Cheever (1850:89) describes the aftermath of whaling operations onboard a whaleship of the 1840s: "The decks, which have hitherto been kept scrupulously clean, are now covered with oil, and it is only by keeping a thick coat of sand scattered over them that the crew are enabled to get about without slipping. The smoke from the try-works blackens every face, so that the watch on deck resembles a party of colliers. Each rope, too, exposed to its influence, is coated with lamp-black, and the clothing of the men saturated with oil. Even the sails, which on the passage were of a snowy whiteness, receive their share of defilement." Thus, all hands were ordered to clear the heavy, black, oily film and make the vessel shipshape (Browne 1850:23; Cheever 1850:89; Dulles 1933:131; Weiss et al. 1974:89). Cleaning involved removing all remaining blood and blubber from the main deck and between decks, then scrubbing all bulwarks, decks, bits, the gangway, the cutting stage, and every other surface using scrub brooms and lye made from the ashes of the blubber scraps, saltwater, and sand (Brown 1887:287; Davis 1874:94; Grant 1932:98; Olmsted 1841:68; Stackpole 1967:51). James Temple Brown describes the nature of sperm oil by stating that "in its natural condition when fresh may be washed off with comparative ease, but after being cooked it removed with difficulty" (Brown 1887:287). The tryworks were scrubbed and the pots cleaned and polished inside until they "shone like silver punchbowls" (Grant 1932:98), then covered with wooden lids to keep debris from contaminating them (Palmer 1959:145). The day after processing was usually given to the crew to wash themselves and their clothes to remove the omnipresent oil and grease, as well as the odor of the fetid smoke of the fires, which all but the old whalers found disagreeable (Browne 1850:51; Grant 1932:98; Hohman 1928:173;

Nordhoff 1874:129–130; Olmsted 1841:68). Except for the upper masts and sails, once the big cleanup was complete, the ship and its crew appeared as they did when they left port (Shapiro 1959:39).

While the foremast hands scrubbed the ship, other tasks were completed in anticipation for the next round of whaling. All implements needed for cutting in and trying out were cleaned and put away. The carpenter and his assistants repaired any damage to the boats incurred through encounters with whales. Occasionally it was necessary to completely replace a stove or smashed boat—referred to by whalers as having been "chawed"—during such encounters. Sometimes called boat surgery, this was done by constructing a new one out of wood shipped onboard for that purpose or assembling them from prefabricated parts (Grant 1932:72, 102). The mates in charge of each boat again supervised their outfitting to ensure that no required gear was overlooked, and the boatsteerer expertly coiled whale line into the tubs (Grant 1932:44). In the meantime, the blacksmith (or cooper if no blacksmith shipped onboard) mended and straightened whalecraft, while helpers incessantly turned the grindstone to ensure that all the various hunting and cutting tools were sharp. Once sufficiently honed, they were sheathed and stored in an easily accessible location (Browne 1850:131; Doane 1987:67; Dulles 1933:110; Grant 1932:38–42; Perkins 1854:25).

With the ship clean and all equipment prepared for whaling activities to begin anew, regular watches resumed. Factors such as seasonal migrations of whales and numerous ships hunting in the same area affected the amount of time between whaling activity. While cruising the grounds, the lookouts often sighted whales on the horizon, and the boats were lowered in pursuit. Despite the best efforts of the whalers, many of those attempts proved fruitless, and they returned to the ship empty-handed (Dulles 1933:155–156). Each time the boats were hoisted back into their cradles, the ship and boats were again readied for the next call from the masthead and for the pelagic whaling process to begin again.

Archaeological Evidence of Preparing the Ship for the Next Round of Whaling

As with the gear associated with stowing down, many of the tools used for preparing the ship for the next round of whaling were manufactured from organic materials and are thus difficult to identify on a shipwreck site. For example, the blocks and brushes used for the seemingly unending chore of scrubbing the oil and gurry from the ship were composed of wood and thick plant fibers (Perkins 1854:29). Evidence of other tasks carried out during this stage, like the sharpening of tools, however, has been found at the *Two Brothers* shipwreck. Discussed previously, the grindstone found on the site is an artifact that was utilized at almost every stage of the cyclical pelagic whaling process. Since most of the industrial activities involved in pelagic whaling resulted in the dulling of tools, turning the grindstone was especially important in preparing for the next round of activity.

The other tasks undertaken at this stage were those carried out by carpenters, coopers, and blacksmiths. These included maintenance duties such as the repair of damaged whaleboats or construction of new ones to replace those lost in the chase and the straightening of bent whalecraft. Since the need for accurate harpoons and well-built, sturdy boats was of utmost importance to a hunt, these tasks were critical for success, and whaleships were equipped with a vast array of specialized tools to ensure that they could be completed quickly and properly. Most outfitting books include sections titled "Carpenter's Tools" and "Cooper's Tools" (NBFPL 1840; NBWM 1832), while others also include lists of "Blacksmith's Tools" or simply "Hardware" (NBFPL 1841, 1845). Among the many different classes of hand tools listed in those books were assorted types and sizes of hammers, mallets, saws, jointers, planes, axes, adzes, squares, compasses, vices, augers, drivers, chisels, irons, files, knives, awls, and gouges, as well as larger items like anvils and bellows (NBFPL 1840, 1841, 1845; NBWM 1832).

Several heavily concreted metal objects were recorded at the *Two Brothers* shipwreck site, and at least two of them are thought to be craftworker's tools. Located in a large pocket in the reef in Section B, these were found along with a grouping of artifacts ranging from galley wares to whalecraft. Among this collection, two classes of tools have been identified: a possible caulking iron, which was a flat, broad implement used for driving oakum (caulking material) between the seams of a vessel's hull planks to make them watertight (Doane 1987:260); and a possible chisel, or a kind of punch that tapered to an edge at one end and was used for shaping materials. Both tools were standard pieces of equipment on whaleships and were used for maintenance. Their location on-site suggests that they were likely kept together on deck (with many other tools) and washed overboard during the wrecking event.

Off Watch

During the period between watches whalers engaged in several activities, many of which were dictated by the space afforded them. During the day, the alternate watch was busy on deck tending to the myriad chores ordered by commanding officers; thus off-duty crew were obliged to spend much of their time in their quarters. Although the practice of recycling merchant ships for use in the whale fishery meant that the hull designs of nineteenth-century whaleships varied, the process of refitting allowed interior spaces to be reconfigured to best allow for maximum cargo capacity. As such, the general layout of a wooden whaleship in the nineteenth century conformed to a standardized pattern that apportioned most of the internal space for oil storage or blubber processing, and the rest was compartmentalized to create accommodation and storage for supplies. Accommodation onboard whaleships was always located at its two ends; average mariners were housed in sparse and cramped conditions in the front of the ship, known as the "fo'c'sle" (forecastle), while officers enjoyed better comforts in the more commodious aft end of the ship. This physical separation reinforced the hierarchy of the chain of command and is reflected in the terms used to refer to the groups—average mariners sailed "before the mast" (Dana 1842), while the ship's officers were often referred to as "the afterguard," since they resided in the after section of the ship (Creighton 1990:537).

The forecastle differed little from vessel to vessel. Located in the extreme forward part of the ship and directly below the main deck, its form was roughly triangular with sides lined with parallel rows of narrow bunks that accommodated 12 to 20 hands (Hohman 1928:126; Perkins 1854:19). Made of ordinary plank timber and cushioned with an uncomfortable mattress of cornhusks referred to as a "donkey's breakfast," the bunk was the only physical space onboard that a whaler had to himself. Thus, some were adorned with curtains of calico cloth to keep them private, were personalized with lithographic images or other mementos, and/or had makeshift shelves for holding a lamp (Ashley 1926:54; Dolin 2007:256; Perkins 1854:19). Each hand shipped aboard with few belongings, usually in either a canvas sack or a sea chest that was lashed to the floor and doubled as a bench. These were generally the only furnishings in the forecastle aside from a small table and lamp (Brown 1887:226; Grant 1932:30; Olmsted 1841:52; Perkins 1854:19). The forecastle was accessed via a small companionway, which also served as the only ventilation and source of natural light (Freeman 1951:240; Grant 1932:30). Due to the combination of poor air circulation and the odors from grimy and wet clothes, oil lanterns, tobacco, unwashed bodies, and the general filth associated with whaling operations, the quarters of average sailors were often described as vile (Brown 1887:226; Dulles 1933:84; Grant 1932:30; Perkins 1854:19). Though attempts were occasionally made to reduce the filth in these quarters by scrubbing them out, generally they were "black and slimy with filth, very small, and as hot as an oven" (Browne 1850:43; Dulles 1933:84; Olmsted 1841:52).

The after part of the whaleship, in contrast, was compartmentalized to create spaces not only for officers but also for storage of ship supplies and food. Accommodation in this section was far better than those of the crew. The captain's quarters were furthest aft and included both a comfortably furnished stateroom on the starboard side—often equipped with a gimbaled bed—and a well-appointed cabin or private office in the center (Brown 1887:226; Whipple 1979:92). Just forward of the captain's cabin were smaller state rooms fitted with ordinary bunks, which were dedicated to the mates; on the larboard side adjacent to the pantry was the first mate's cabin and forward of it were bunks of the third and fourth mates, while on the starboard side in the same position were the second mate's cabin (Brown 1887:226; Hohman 1928:125; Whipple 1979:92). Farther forward of the officer's staterooms and separated from them, was the area known as steerage. This small, poorly ventilated, and dimly lit area contained eight bunks and accommodated the more skilled members of the crew including the boatsteerers, the carpenter, the cooper, the blacksmith, and the steward (Hohman 1928:126; Shapiro 1959:24). Though the quarters of those in steerage were spartan compared to those of the officers, they were better than those of the foremast hands; as historian Alan Mawer suggests, "if the forecastle was the ship's hell hole, steerage was its purgatory" (Mawer 1999:174).

If the weather and conditions permitted, whaling crews spent the free evening hours on the foredeck where they sang, danced, spun yarns, and relaxed (Browne 1850:46). For the most part, however, their spare time was in the forecastle where they could escape the watchful eyes of the officers (Martin 1974:36). There they could relax through reading, writing letters, keeping journals, sketching, and mending worn clothes. Or if rest was not desired, they could indulge in less personally enriching pursuits such as gambling and fighting, provided the captain did not find out (Browne 1850:111; Delano 1846:28; Martin 1974:36; Olmsted 1841:52; Verrill 1916:190). Except during the night, the quarters of the foremast hands were generally loud, active spaces where crew members chattered away, laughed, played music, sang ditties, and spun yarns (Browne 1850:43; Dulles 1933:84; Perkins 1854:19). All the while they smoked incessantly—tobacco being the "whalemen's narcotic" (Dolin 2007:260; Mawer 1999:174–175; Whipple 1979:78)—and many spent idle hours carving knickknacks and ornamental items out of shells, whale teeth, and whalebone (Brown 1887:231; Dodge 1882:15; Mawer 1999:174–175; Whipple 1979:126). Generally known as scrimshaw or scrimshander, carving was the favorite pastime of whalers who used little more than a jackknife to create crafts ranging from polished whales' teeth with intricate artistic scenes etched onto them, to more functional pieces such as tool handles, pie jagging wheels, yarn swifts, and walking canes (Sturtevant 1955:7–10; Whipple 1979:126–127). So pervasive was this practice that it became a sign of prowess among the average whaler (WPA 1938:15).

Off-watch time was undoubtedly spent differently by the officers and boatsteerers. Though information pertaining to officers during these periods is scant, it is presumed that the duties and responsibilities of their position kept them busy, and fraternizing with the foremast hands was not tolerated. Since the individuals holding these positions aspired to rise through the ranks, it is likely that much of their free time was devoted to the study of navigation and sailing, as well as to general personal enrichment through reading and writing.

One pastime activity enjoyed by officers and crew alike was fishing. When in warm latitudes, fish, sharks, dolphins, or porpoises could often be found off the bow, and if conditions permitted, attempts were made to catch them using hooks baited with white rags or spearing them with harpoons. When successful, the products of these endeavors were taken to the cook to be fried, and they always made a welcome change to an otherwise monotonous menu (Brown 1887:228; Davis 1874:123–124; Davis et al. 1997:251; Dickson 2011:173; Dodge 1882:15; Dolin 2007:260; Grant 1932:50; Whipple 1979:78). The importance of accommodating these activities can be found in outfitting records in the form of fish hooks and lines of various sizes and lengths being consistently shipped onboard from as early as 1765 (Gale Huntington Research Library 1765; Kirby 1860; NBFPL 1845; NBWM 1819, 1832; NHA 1807).

Archaeological Evidence of Off-Watch Activities

None of the artifacts identified at either the *Two Brothers* or *Parker* shipwreck sites are considered directly related to off-watch activities of the crew. The lack of such evidence likely results from the fact that the personal effects of these individuals were few and for the most part organic. For instance, due to the cramped conditions of the forecastle, the belongings that a crew member might bring onboard were limited to what would fit into a canvas sack or a sea chest. Since whalers were responsible for being prepared for all conditions, much of that space was for clothing that provided protection from harsh conditions. Though artifacts identified as personal belongings of sailors such as jackknives or clay pipes have been found on the wrecks of merchant ships, they have eluded researchers on these whaleship wreck sites. The absence of these types of artifacts likely results from the environmental conditions experienced in PMNM; thin metal pieces such as knife blades generally deteriorate in the warm and turbulent waters of the North Central Pacific, while pipe stems and bowls are almost impossible to recognize on the seabed since it consists mainly of dead and broken coral resting on a marl substrate.

Crew Maintenance

Mealtimes onboard early nineteenth-century whaleships were one of the few parts of the day that whalers anticipated. Though the food was generally considered to be less than appetizing, the opportunity to eat was a welcome break from the routine of watches and the boredom that came with the wait for whale sightings (Browne 1850:28). At the outset of the cruise, the captain announced that hard work would be rewarded with plenty of "hot chow" and crews would be satisfied with the portions. Many captains, however, followed the orders of frugal owners and allowed cooks to use only minimal provisions (Browne 1850:37; Dolin 2007:260). Meals generally consisted of only a few ingredients and, unless the ship stopped at an island or port to reprovision and recruit, were basically the same each day. Yet the simplicity of the ingredients was often not the issue; since few whaleships hired experienced cooks, the fare was generally thrown together in a haphazard fashion with little concern for flavor (Dulles 1933:87; Hohman 1928:130). Regardless of quality or appeal, meals provided sustenance, helped crew members avoid scurvy and, for the most part, kept them healthy and ready for action.

Since the food served onboard whaleships was not necessarily made with taste in mind, the desirable characteristics of food supplies were that they could be packed into casks and preserved for long periods of time. Shipped onboard in bulk were large quantities of salted beef and pork (commonly called salt horse or salt junk), flour, premade biscuits called hard tack, rice, legumes, dried vegetables, and dried fruits (Brown 1887:227; Dolin 2007:259). Freshness of the provisions dwindled with the amount of time spent at sea, and exposure to tropical climates usually resulted in at least some spoilage. Little could be done, however, to prevent this, and cooks tried to make it edible (Dolin 2007:260). Though few meals could be considered treats, a few nights each week more desirable fare was prepared. These meals included a spiced, savory dish known as "scouse" and a sweetened dish with dried fruit and molasses known as "duff" (Brown 1887:227; Dolin 2007:259–260; Dulles 1933:87; Mawer 1999:168–169; WPA 1938:17).

Although the quality of food served on whaleships was considered poor by many whalers, others suggested that it was far better than that served onboard merchant ships of the period (Dulles 1933:87). Certainly, the crews of schooners and brigs employed for Atlantic voyages would have offered few complaints. Since those cruises generally lasted less than a year, fresh provisions were abundant, and meals were followed with plum duff—hence those vessels operated in the "plum pudding fleet" (Brown 1887:233; Mawer 1999:201). Though the crews of the larger Pacific-bound ships were not as fortunate, outfitting lists indicate that the number of ingredients available to the cook for enhancing flavor increased greatly in the early nineteenth century. For instance, an outfitting book dating to 1807 shows that along with dried fruits and pickles, sweeteners such molasses, sugar, and chocolate were shipped onboard (NHA 1807). By the early 1830s, the menu of items expanded to include spices like mustard seed, black pepper, cayenne pepper, ginger, allspice, nutmegs, cloves, cinnamon, sweet oil, pepper sauce, lemon syrup, sage, and coarse salt (NBWM 1832). Though most of those ingredients were likely intended for preparing the meals of officers, the large quantities of some items suggest that the crew benefited from them as well.

Provisions were tightly packed and sealed in casks of varying size. For instance, large barrels were used for keeping staple foodstuffs since they were shipped in great quantities, while much smaller casks were needed to store specialty items. In many cases, early to mid-nineteenth-century whaleship outfitting books did not include the names or sizes for the casks used for such purposes; instead, contents were generally measured in gallons (NBFPL 1840, 1841, 1845; NBWM 1832;). Once onboard, stowage of provisions depended on immediate need; some of the casks containing food were specifically placed in spots easily accessible to the cook, while the others were stored in the upper tier of the forehold to prevent them from being contaminated by oil seeping from casks (Brown 1887:227; Howard 1996:441). Once opened, the contents of a cask of provisions were kept with other daily use items in the pantry, which was a secured compartment located on the lower deck near the officer's quarters.

All meals were prepared in the ship's designated kitchen area known as the galley, which was located between the mainmast and the mizzenmast. Although on later ships the structure that enclosed this area was more substantial, early nineteenth-century whaleship galleys were little more than simple wooden shelters. Referred to by whaler Francis Allyn Olmsted as a "little kennel large enough for the cook and his stove" (Olmsted 1841:51), this tiny space was where the cook kept cauldrons and pans—collectively referred to as "coppers"—as well as all other equipment used to prepare meals for all onboard (Brown 1887:227; Draper 2001:30; Olmsted 1841:51). According to various outfitting books, the cooking utensils used in the preparation of meals included "cooks pots," saucepans, stew pans, and frying pans, as well as ladles, carving knives and forks, chopping knives, and skimmers (NBFPL 1840, 1845; NBWM 1832).

Meals were served at regular times each day unless whaling operations were underway. Breakfast was served at 7:00 in the morning, lunch at noon, and dinner at 5:00 each evening (Brown 1887:228). Since the hierarchy of command was directly reflected in the quality of food served and in dining conditions, the commissary department was divided into three sections: one for the officers, one for those in steerage, and one for the foremast hands (Dolin 2007:259; Hohman 1928:130). The captain and officers took meals together in the cabin and were given the best-quality food and condiments served on place settings of porcelain dishes and silverware (Perkins 1854:19; Shapiro 1959:24). Some captains allowed the boatsteerers to dine in the cabin after the officers were finished. Though the food was much the same as that of the officers, often boatsteerers were required to eat in steerage (Brown 1887:228; Olmsted 1841:52).

The dining experience for the afterguard afforded as many of the comforts of home as possible for a working ship. The cabin was well ventilated, illuminated via a large skylight, and housed a hardwood dining table fastened to the deck (Hegarty 1960:79). The ship was equipped with a range of tableware and serving dishes on which meals were formally served by the steward (Hohman 1928:133). Preprinted outfitting books dating from the 1830s to the 1860s contain a standard list of dishes and condiment containers intended for the cabin under the subheading "Crockery." Pottery types include shoal plates, soup plates, small plates, pudding dishes, oval dishes, gravy dishes, soup tureens, sugar bowls, butter tubs, salt dishes, castor bottles, tumblers, wine glasses, decanters, mugs, cups, saucers, bowls, covered dishes, pitchers, and platters (Kirby 1860:31; NBFPL 1845; NBWM 1832).

As for crew meals, when ready the bell rang to indicate the time and associated meal (e.g., "seven bells" was breakfast), and the hands came aft to collect it. Regardless of its quality, they generally indulged in the treasured privilege of making "uncomplimentary remarks about the cook and all his ancestors" (Grant 1932:48; Perkins 1854:29). Depending on the weather, the crew took their meals either on the foredeck or in the forecastle using sheath knives, mugs, and small wooden tubs called "kids" (Grant 1932:48; Perkins 1854:19; Shapiro 1959:24). If a foremast hand, however, preferred to use proper eating utensils, they could be purchased from the slop chest; records pertaining to the whaleship *Condor* list iron spoons, tin pots, and tin pans among the items available (NBWM 1832). Regardless of what dishes they used, crew members were responsible for washing them and cleaning up after meals (Brown 1887:227).

One of the main concerns regarding provisioning and meals was the constant threat from scurvy (Freeman 1951:240). Scurvy is a potentially fatal malady caused by a lack of essential vitamins and is often associated with long sea voyages (Hohman 1928:138; Nash 2001:34). Symptoms include sore gums, bleeding, extreme fatigue, foul breath, and swollen limbs; depending on the health of the crew at the start of the cruise, these can manifest in as little as six weeks and eventually cause circulatory failure (Dolin 2007:261–262; Nash 2001:34). The main defense against the disease is adding foods high in vitamin C to the diet. As such, among the fruits and vegetables shipped onboard during outfitting were large quantities of potatoes and onions, which were known to preserve better than other vegetables (Hohman 1928:138; Mawer 1999:170). Since sick crews generally meant poor outcomes when whaling, visits to ports or friendly islands were made to obtain fresh supplies when necessary or if a convenient opportunity presented itself (Mawer 1999:170). Though many of the foods found in tropical regions helped to keep scurvy at bay, pure lime juice was revered as the best "anti-scorbutic" (Gifford 1998:142; Hohman 1928:138–39).

If a ship was offshore for an extended period and fresh provisions ran low, some captains elected to give rations of rum to stave off scurvy (Dodge 1882:21). In the early years of pelagic whaling, large quantities of rum and other types of alcohol were commonly included in the outfits of all whaleships, and these were served from time to time as grog rations (Brown 1887:228). The presence of alcohol onboard New Bedford vessels is evidenced by records relating to the inaugural voyage of the whaleship *William Rotch* in 1819. These documents indicate that a relatively large quantity of liquor was shipped onboard including "132 gallons and four bottles of rum," "gin case and barrels" (quantities unlisted), an undisclosed amount of "Lisbon wine," and an undisclosed amount of brandy (NBWM 1819).

The temperance movement of the 1830s, however, resulted in many captains forbidding alcohol on board ships (Brown 1887:228; Dulles 1933:88; Hohman 1928:136). This sentiment grew quickly, and after 1840, it was only allowed for medicinal purposes (Brown 1887:228; Littlefield 1906:10; Mawer 1999:138). Outfitting records from the 1830s to the 1860s indicate that the only officially sanctioned alcohol onboard were meager amounts of "N.E. Rum, Holland Gin, Brandy, and Port Wine," listed under the subheadings of "For Medical Purposes," "Medical," or "Medicinal" (Kirby 1860; NBFPL 1841, 1845, 1851; NBWM 1832). Still, alcohol could be found on ships of this period. Spirits were sometimes shipped as a trade item, and in some cases, small amounts of it were also brought onboard by the captain and officers for personal consumption (Brown 1887:227). Though alcohol and all other intoxicants were forbidden for crews (Dulles 1933:88–89) and bootlegging was an offense of ship rules by the middle of the nineteenth century, it was often smuggled onboard by many of the foremast hands (Hohman 1928:136). Whether doled out as grog rations to motivate crews on particularly wet and cold days (Delano 1846:24), served with meals to the afterguard, used for medicinal assistance, or snuck onboard as contraband, various types of liquors and wines would certainly have been found onboard early nineteenth-century American whaleships.

The rough and filthy living conditions on whaleships, the long periods of subsistence on highly salty diets, and the environmental extremes experienced by whalers all contributed to the range of maladies onboard. Common sicknesses included tropical fevers, dysentery, venereal disease, rheumatism, tetanus, tuberculosis, pneumonia, colds, and depression (Dolin 2007:261; Dulles 1933:88). While British ships of this period were legally bound to employ trained doctors for cruises, health onboard American vessels was overseen by captains who were utterly untrained in medicine (Grant 1932:97; Hohman 1928:137; Mawer 1999:177). When acting as a physician, a captain was guided by the contents of the medicine chest and an accompanying booklet of instructions that was legally required to be carried onboard (Dolin 2007:262; Hohman 1928:137; Poole 1977:3).

Sometimes called "doctor boxes" (NHA 1807), these standardized kits contained a limited range of drugs and medicines housed in small bottles, as well as a few utensils such as mortars and pestles, balances, weights, and measuring devices (Lipman and Osborne 1969:124–125). Other items at the captain's disposal were bandages and splints for setting broken bones, a few surgical instruments, and liquor, which was often used as anesthetic (Hohman 1928:137; Lipman and Osborne 1969:124–125; Littlefield 1906:10). While these medical supplies helped to ease suffering and provide relief from pain, some afflictions and injuries were far more serious. In such cases, the captains often set a direct course for a port where proper medical assistance could be obtained or the crew member could be discharged to recuperate. If an ailing crew member passed away onboard the ship, it was customary to bury them at sea in a ceremony presided over by the captain and attended by the ship's company (Paddack 1893:87).

Aiding a sick crew member was just one of the causes that led captains to suspend whaling and make port. Other main reasons included taking shelter from seasonal storms; refitting and repairing ships and rigging; transshipping cargos to homebound vessels; restocking supplies of fresh fruits, vegetables, firewood, and water; and recruiting replacements for crew members who had perished, deserted, or were discharged for legal issues. Whatever the initial reason, consideration of all other maintenance needs was given any time a ship came into port since fees for doing so were charged from a very early period (Birkett 2000:76; Browne 1850; Kuykendall 1934:367). Thus, full advantage would be taken of any visit to an established port, though no time was generally wasted for social purposes unless the ship was wintering over (Mawer 1999:138).

Archaeological Evidence of Crew Maintenance

Archaeological evidence of the casks used for food storage on *Two Brothers* and *Parker* is impossible to distinguish. The subtropical marine environment of PMNM is inhospitable to wooden remains, and, as with the hull of the ship itself, all casks onboard would have been subject to decay caused by marine organisms soon after submergence. With the wooden components gone, all that remained were iron hoops used for binding. Though it is plausible that hoops found on shipwreck sites could be used to extrapolate the cask size, the thin iron from which they were constructed generally becomes brittle and breaks down in seawater. Thus, although pieces of several iron hoops were found at both sites, it is impossible to determine what materials were stored in them or if they were unused and stowed during the wrecking. While it is estimated that thousands of individual hoops would have been shipped onboard a nineteenth-century whaleship, an actual count is difficult to determine since they were generally purchased in bulk. For example, the outfitting book for the New Bedford whaleship *Condor's* 1832 cruise simply lists two and a half tons of "hoop irons" onboard (NBWM 1832).

Aside from cask hoops, the only other example of a food storage container identified at either wreck site is an intact stoneware jar. Located in Section B of the *Two Brothers* site near other artifacts associated with a whaleship's aft end, this utilitarian vessel was likely used in the ship's galley for storing daily use items. Generically referred to as a "ginger jar" since its shape somewhat resembles Chinese porcelain vessels of that type, the salt-glazed earthenware container is made from a gray paste, and its glaze has a yellowish appearance. Because the jar is firmly concreted into the reef, only half of it could be inspected; however, no signs of decorations or maker's marks were noted on any visible surface. Although stoneware pottery fragments of various shapes are common among artifact assemblages of most archaeological sites dating to this period, the ginger jar identified at *Two Brothers* is the only known intact jar of this type documented in a shipwreck context.

Four heavily concreted cast-iron cooking pots were recorded among the remains of *Two Brothers*. These were all located near one another in a large pocket in the reef in Section B of the site, which is consistent with historic descriptions of the galley being in the after part of a whaleship (Olmsted 1841:51). Although there is slight variation in the sizes of these pots, they are all of similar design and shape (Figure 6.11). Features include a bulbous body that narrowed toward the top, a flaring collar neck with two earlike handles attached to the rim at opposing sides, decorative "cordons" embossed around the body's exterior, and three small legs on the bottom (Franklin 2005:139; Hume 1991:175–76; Neumann 1984:176). Ivor Noël Hume (1991) suggests that cooking vessels of this basic form were made for centuries using cupreous materials, but by the seventeenth century, the dangers of copper poisoning led to a shift to iron production (Hume 1991:176). Although cast-iron stoves were popular onboard many ships from the latter part of the eighteenth century, there is little indication that whaleships were equipped with them until at least the mid-nineteenth century.

Similar artifacts have been identified on numerous mid- to late eighteenth-century shipwreck sites, including the gondola *Philadelphia*, sunk in Lake Champlain in 1776 (Bratten 2002); the British transport *Industry*, lost at

Figure 6.11. Photograph postconservation of one of four cooking cauldrons documented at Section B of the *Two Brothers* shipwreck site. (Courtesy of Papahanaumokuakea Marine National Monument.)

St Augustine, Florida, in 1764 (Franklin 2005); and the Storm Wreck, lost at St. Augustine, Florida, around 1780 (Meide et al. 2011). One of the four pots recovered from the *Two Brothers* site was conserved for eventual display. During conservation, a possible mark resembling the letter M was noted on the pot's inner rim (Fox 2012:4). Although the meaning of this mark remains unknown, the practice of marking cauldrons was not uncommon; according to Hume, nineteenth-century cauldrons were frequently embossed with capacity numbers in Arabic figures (Hume 1991:176).

A number of ceramic sherds thought to be associated with the aft dining experience were identified at the wreck of *Two Brothers*. Sherds of shell-edged green and shell-edged blue pearlwares, a pearlware mug handle, and molded pearlware sherds were all documented at the site. The most diagnostic of the ceramics recorded are the shell-edged pearlware sherds (Figure 6.12). Pearlware ceramics are British earthenware that were first introduced by Josiah Wedgwood in the 1770s. Wedgwood modified the already popular creamware using a blue-tinged glaze, a combination that produced a whiter ceramic that more closely resembled Chinese porcelains (Moran 1976:198; Sussman 1977:105). Pearlware achieved great popularity in the late eighteenth century and became the dominant ceramic ware until its decline around 1820

(Ford et al. 2008:85; Hume 1991:129–130). Although used to produce a wide range of ceramic forms, shell-edged plates with blue or green painted rims were the most common (Hume 1991:129–130). The decorative edges of pearlware plates changed over the course of their production; on early examples the molded relief on the edges takes the appearance of intricate ruffles, while later versions are simply impressed curved lines with both regular and irregular spacing patterns (Sussman 1977:106). The rims of each of the blue and green shell-edged pearlware sherds identified at the *Two Brothers* site exhibit the latter pattern. The outer edges of both are scalloped and have closely spaced vertical impressions of varying length to which pigment was added; this style was common on both eighteenth- and nineteenth-century pearlware plates (Sussman 1977:107).

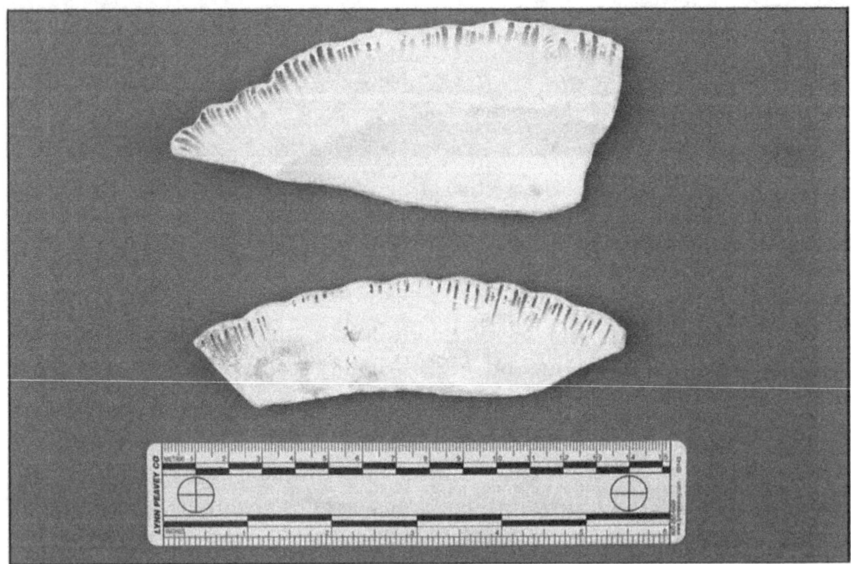

Figure 6.12. Photograph of postconservation green and blue shell-edged pearlware sherds documented at Section B of the *Two Brothers* shipwreck site. (Courtesy of Papahanaumokuakea Marine National Monument.)

Fragments of at least two glass bottle types provide archaeological evidence of alcohol usage onboard *Two Brothers*. Although the large quantities of alcohol often shipped on whaleships of this period would have likely been stored in wooden kegs, smaller and more manageable containers like bottles would have made serving easier. The remains of one of the two bottle types identified at the wreck site include numerous green glass shards and the bases of dark olive green glass bottles. Found scattered around Section B, though most of the shards are very small, two are large enough to determine the part of the bottle that they comprised: one is a section of the cylindrical body, and the other is part of a shoulder. A round bottle base was recorded in each of the two sections of the site. Though both of these were heavily coated with marine encrustation, their colors and some features were discernible: each is completely round in shape and approximately 9 cm in diameter, each has a "kick up" that rises roughly 6 cm in the center, and each appears to be made of dark olive green glass (also referred to as "black glass" for its dark hue) (Lindsey 2023). These are most likely the bases of "tall, moderately slender bulged neck spirits/utility cylinder bottles," which were first produced in the late eighteenth century but became highly popular for holding different types of liquids by the 1820s (Lindsey 2023).

The other type of alcohol bottle identified at *Two Brothers* is a "case bottle." Distinguishable by its nearly flat base, square sides that taper in from the shoulder down, and short neck with a flared lip, case bottles were designed to fit together in a packing box known as a case or "cellar" (Hume 1991:62; Lindsey 2023). Although the shape and style of these bottles originated in seventeenth-century Europe, versions were made in the United States by at least the early 1800s (Lindsey 2023). Case bottles are commonly referred to as "Dutch gin bottles" due to their primary purpose of shipping European-style gin known as "Hollands" or "Geneva" in the late eighteenth century (Hume 1991:62).

The bases of two case bottles were identified in proximity to one another on the northern reef flat in Section A of the site (Figure 6.13). Each of these measures approximately 8 cm squared, appears to be dark olive green in color, and exhibits a diagnostic round scar on its base. According to bottle historian Cecil Muncy, scars such as this are the result of a pontil rod being broken off during manufacture and indicates that the bottle was made prior to the mid-nineteenth century (Munsey 1970:85). The presence of these artifacts at the wreck of a pelagic whaleship is not uncommon since "Holland gin" was among the four main types of medicinal liquors listed in the outfitting records of early to mid-nineteenth-century whaleships (Kirby 1860; NBFPL 1841, 1845, 1851; NBWM 1832).

Homeward Bound

Most early nineteenth-century whaling cruises lasted several years and were characterized by contrasting periods of idleness and intense activity. The advent of transshipping oil cargos at established ports resulted in voyage lengths becoming ever longer. This practice allowed the ships to continue whaling for successive seasons and to make the most of their time around Cape Horn. Eventually, however, it was necessary to return home, and the decision

Figure 6.13. Diver measuring the base of a case bottle documented at Section A of the *Two Brothers* shipwreck site. (Courtesy of Papahanaumokuakea Marine National Monument.)

regarding when that occurred rested solely with the captain. Whether cargos were transshipped or not, it was only when the captain was satisfied that an acceptable quantity of oil had been obtained that he gave the order to begin the homeward passage (Grant 1932:124).

Since no opportunities were wasted while a ship transited, operations continued as usual and masthead lookouts constantly maintained for much of this time. Even if the spaces for storing oil were full, a good whaling captain never passed on the chance to obtain more; thus, if whales were spotted, the boats were lowered in pursuit. It was only when a ship was in the Atlantic that preparations were made for concluding the voyage. Since most captains of the period took pride in their vessel's appearance upon its return, several chores were undertaken to ensure that it looked its best when entering port (Brown 1887:234). These included deeply scrubbing the decks with lye and sand until they looked new, painting the outside of the ship by hanging over the side in a sling attached to special rigging, scraping the masts, overhauling the rigging, and sometimes breaking out and bending a new suit of sails (Brown 1887:234; Davis 1874:391). When nearing the New England coast, crews could be certain that whaling operations were at last complete when the time-honored tradition of knocking down was ordered and the tryworks dismantled (Ashley 1926:96; Leavitt 1973:10). Furthermore, all the boat equipment and accessories were cleaned and stowed in marked casks, the cutting tackles and pendants taken down, and all whaling and processing implements bundled in canvas and stowed away (Brown 1887:230, 288).

As the ship neared the shore and land was signaled, crews were overwhelmed with joy and anticipation (Browne 1850:480–481). Once the ship was safely anchored in port, the crew was required to await a shipkeeper to take over its watch. Before leaving the ship, the crew went below to "take a glass of grog," as was customary in the early nineteenth century (Delano 1846:65). After the crews disembarked, they remained in port until they received their lays, at which time they scattered like the winds, many of them to never go to sea again.

Conclusion

From the moment the mooring lines were cast off and a ship departed its home port, it morphed into a microcosm of industrial life with one goal in mind: catching whales and filling the hold with their produce as quickly and efficiently as possible. To achieve this, a hierarchical labor system bound to a strictly regimented schedule of routine daily tasks was critical. While activities like sighting, harpooning, and processing cetaceans in the name of profit are often romanticized in popular literature, mundane chores such as sharpening whalecraft and maintaining the health and morale of the crew were of no less importance. Keeping the vessel, equipment, and labor force shipshape and ready allowed it to transform into a floating production platform that operated with maximum efficiency. But despite the best efforts of its officers and crew, ships were at the mercy of an environment that regularly dealt conditions that proved disastrous. As such, the remains of whaleships lost in the constant search for profit offer a glimpse into the lives of the seafarers who lived and toiled in these maritime industrial workplaces. When considered in this way, it is easy to see that the often-glorified actions encompassed in the cruise operations stage were simply one component of the larger American whaling tradition.

Conclusion

Like all commercial sailing vessels of the early to mid-nineteenth century, pelagic whaleships provided a platform upon which work was carried out in a structured environment and under the supervision of a strictly defined labor hierarchy. Whaleships, however, differed from other vessels of the period in that they were also outfitted with the equipment needed to chase and capture whales, remove and process blubber, and store large quantities of oil until it could be shipped or transshipped to ports in New England. As they were essentially floating industrial sites, these vessels allow for the exploration of themes such as organizational structure, conditions experienced by crew, methods and equipment used in whale hunting, operational scope, and environmental conditions. The following discussion considers the value of exploring the remains of whaleships as industrial sites to better understand the American whaling experience during its golden age. Through the adaptation of interpretative methods from one subdiscipline to another, connective themes are illuminated. The technological and organizational perspectives gleaned from integration of datasets acquired through archival research, archaeological investigation, and comparative studies of museum collections provide a more nuanced understanding of the social factors that influenced these industrial workplaces.

Expanding the Horizon

Early nineteenth-century American pelagic whaling was undoubtedly an industrial activity based on the extraction of resources from the marine environment. Success in this industry depended on adherence to a developed industrial system that involved considerable planning and strategic decision-making, specialized tools and processes, and a rigid labor hierarchy. The industrial nature of the material culture assemblages associated with wrecked whaleships distinguishes them from vessels used in other nautical activities of the period. As such, the adaptation of themes commonly used for analysis of industrial archaeological sites offers a different perspective to the interpretation of shipwrecks.

Although some authors have referred to whaleships of this period as "floating factories" (Allen 1973:159; Day 1966:132; Weiss et al. 1974:79;), early nineteenth-century whaleships are more closely aligned to some terrestrial sites associated with primary resource extraction. Much like the whaleships that undertook extended voyages into the largely unknown waters of the Pacific Ocean, remote mining operations on the American frontier were microcosms of interrelated production activities isolated for often long periods of time. To succeed, both operations required a great deal of planning and outfitting, exploration of large geographic areas to find sources of raw materials, and technical knowledge and experience on the part of managers and crew. Due to the organizational and operational similarities between the two industries, some of the interpretative themes used to explore frontier mining workplaces were adapted to interpret pelagic whaleships as sites of resource extraction.

The ships that provided the platform for pelagic whaling activities can be seen as the main physical component of this maritime resource extraction industry. Although the designs and sizes of whaleships evolved with the expansion of pelagic whaling from 1750 to 1815, their hull characteristics changed little over the period of the industry's golden age. Despite modifications to the designs of the ships involved in other commercial enterprises of this period, which led to greater speed and efficiency, such results were of little importance to whaleship owners. Instead, whalers preferred hulls that offered stable working platforms, were easily arranged for industrial operations, and included maximum cargo capacity. Since these same features were also found in vessels designed for shipping bulk cargos and passengers, hulls originally built for such activities were often recycled for use in pelagic whaling. This practice lessened initial investment costs and in turn produced greater financial returns. Though it is true that by the 1830s many ships were constructed specifically for whaling operations, the hulls of purpose-built whalers were designed with the same characteristics.

Unlike frontier mining sites, the remains of whaleships are not characterized by the structures that supported them. While many of the metal fittings used to fasten ship's timbers or attach components of rigging are present, for the most part, the archaeological assemblages of early nineteenth-century pelagic whaleships lost in oceanic environments are typified by the presence of specialized equipment for resource extraction. American pelagic whaling was made possible through the adaptation of methods and tools used in shore whaling to suit the confines of a ship and the demands of seaborne operations. The processes employed and their associated equipment changed little in the first half of the nineteenth century. Among the most distinctive of those tools were the sleek whaleboats and honed whalecraft needed for pursuing and capturing whales; the windlass, specialized rigging implements, and razor-sharp hand tools for stripping blubber from a carcass and preparing it for processing; the bricks, pots, and various fittings that comprised the tryworks used for rendering blubber into oil; the casks, rigging implements, and hoses used to transfer and store oil; and the tools used by crew members to maintain constant readiness for the next chase. These implements were crucial to the success of the industrial process of pelagic whaling, and their presence at a shipwreck site is an immediate indicator of involvement in the trade.

The geographic locations of shipwrecks in remote parts of the Pacific Ocean illustrate the broad industrial seascape associated with American whaling. If pelagic whaleships are to be considered sites associated with resource extraction, is it important to understand how American whalers used knowledge of whale behavior to identify potential hunting areas. For extractive industries like frontier mining, desired resources manifested as discrete ore deposits scattered across large geographic areas. Found through extensive exploration activities, identified veins of ore became the focus of intense industrial activity that impacted the local environment. Pelagic whaling operated in a similar fashion, as it required extensive exploration to find resources scattered across expansive geographic regions, and once located, they were exploited to the maximum possible extent.

Information pertaining to the industrial seascape of pelagic whaling was a crucial component of the system. From the time that the first whaleships rounded Cape Horn and explored the Pacific, masters of vessels compiled information about the physical environment. Since the immense numbers of sperm whales encountered in early voyages were seen as a harbinger for a shift in the fishery's geographic focus, recording such data was valuable for planning subsequent expeditions. The types of information considered most useful pertained to areas where large populations of whales could be found, the locations of islands offering safe anchorages and sources of fresh provisions, and the positions of geographical features that posed potential threats.

As whaling captains learned while fishing in the Atlantic Ocean, sperm whales are a migratory species with predictable behavior. Exploration efforts indicated that whales followed the abundant food sources of warmer environments. In the Pacific, this meant that sperm whales spent the cooler months in southern and equatorial zones but significantly expanded their range to the north during warmer months. Over time, general areas where whales could be found in large concentrations were designated as specific hunting grounds. Although the exact boundaries of these areas were fluid, their relative locations and data pertaining to the best season to fish them were recorded in journals, logs, and on charts carried onboard ships. Unlike other marine resource extraction industries of this period that attempted to keep new hunting areas secret, the whaling fraternity shared information about newly identified zones. In the early years of the Pacific fishery this was conducted by word of mouth, but over time, the information found its way onto charts of the region. Thus, data pertaining to the operational areas for pelagic whaling were an ever growing body of knowledge that was freely provided through both oral and written sources.

The earliest American ships to round Cape Horn found whales in abundance and quickly filled their holds before making the journey back to New England. As news of this discovery diffused, ships quickly followed, and in a relatively short time, whalers noticed a depletion of the whale population on the initial ground. This information led vessels to venture farther from shore in search of the next rich spot. As occurred in the Atlantic Ocean in the mid- to late eighteenth century, the ever increasing number of ships led to a pattern: intense fishing depleted resources on new grounds, which then led to the search for another densely populated area. By the early 1840s, whaleships surveyed much of the previously unexplored expanse of the Pacific and pushed as far north as the Bering and Chukchi Seas, where they found a new species to exploit. Thus, the search for new hunting areas and quick returns expanded the operational range of whaleships, which in turn extended the average length of cruises and had a direct effect on the sizes and rigs of the vessels employed.

Information concerning geographic features and oceanographic conditions was also of great interest to early whaling captains. While pushing the boundaries of known grounds, whaleships encountered many previously uncharted islands, atolls, reefs, and shoals. In some cases, these manifested as verdant island groups that presented potential points for ships to reprovision with food and water. In other instances, however, newly identified areas posed the threat of shipwreck due to extensive and mostly submerged reef systems that were often unnoticeable from a distance. Thus, when uncharted geographic features were detected, it was of the utmost importance that their locations were recorded as accurately as possible. Since this could only be done with the surveying technology available at the time, reported locations were often only estimates. As the use of chronometers increased so did the accuracy of those positions, which in turn reduced the risk of shipwreck.

Although many of the early discoveries amounted to little more than a captain taking a sighting from a distance and ascribing a name to the feature, others were cautiously approached and examined. When possible, geographic position data was augmented with comments regarding physical information such as water depth obtained from lead line measurements, hydrographic data pertaining to currents and tides, and/or data regarding weather experienced.

Any of this information would have been relished by whaleship owners and captains of the early nineteenth century since it assisted with safe operations and helped with voyage planning.

In terms of the cultural geography of the region, the lasting effects of these early exploratory voyages and sightings can be seen in place names located throughout the Pacific Ocean. While most of these features were named, inhabited, or at least known to Indigenous seafarers, when encountered by American or European whalers (either sighting or wrecking), their positions were generally recorded in logs or journals using titles derived from the names of ships, captains, or home ports. In some instances captains endeavored to document Indigenous names of geographical features and integrated them on their charts. For many others, however, the titles attributed to encounters with whalers linger despite modern efforts to revert to Indigenous names.

Though the expanse of the Pacific was largely unknown, it represented a rich industrial operations area for pelagic whalers, and through sharing information, they greatly increased the chances for financial success. By combining data about the seasonal movements of whales between known grounds, whalers hunted throughout the year to obtain the greatest returns in the shortest amount of time. By using shared knowledge about areas of potential hazard, captains eliminated many of the otherwise unknown risks until federal mapping expeditions officially charted them. Thus, the various hunting grounds located around the Pacific region were inextricably tied to the larger industrial seascape of marine resource extraction, and the effects they had on the environment can be seen through diminished whale populations.

As illustrated in the preceding discussion, early nineteenth-century American whaleships were floating industrial sites, so analyzing their remains as such helps to better understand the many interrelated, and often unconsidered, facets of the trade. Organizational and operational similarities with frontier mining activities in the American West provides themes that allow for the investigation of pelagic whaling of the period as a maritime resource extraction industry operating within an enormous industrial seascape. The integration of archaeological and historical data relating to those themes illustrates that not all industrial elements are visible in the archaeological record. As such, this approach not only contextualizes the material culture remains of ships by exploring the methods and tools used in the various stages of extracting oil; it also helps to understand how shipboard processes functioned within the larger industrial system of pelagic whaling. Historical data pertaining to the vast amount of preparation required, as well as the shared knowledge of the physical environment and the behaviors of whales, proved to be equally as important to industrial success as the physical characteristics of the ships that provided the platform for those processes.

Standardizing Practices

The maritime industrial workplace that existed onboard whaleships of the early nineteenth century was characterized by physical and organizational features that developed out of mid-eighteenth-century technological innovation. The installation of the shipboard tryworks around 1750 resulted in unprecedented growth of the American whaling enterprise, which in turn led to standardization of practices and technologies to create an exceedingly effective industrial system. Although some refinement of the technologies employed onboard American vessels occurred in the decades that followed, throughout much of the industry's golden age, the workplaces that existed onboard American whaleships adhered to this basic structure.

From a physical standpoint, early nineteenth-century pelagic whaling workplaces were characterized primarily by the vessels employed, since they provided the platform for operations. The integration of shipboard tryworks saw the size and style of whaling vessels grow from roughly 60-ton sloops to full-rigged ships of 200 to 400 tons. Once equipped for blubber processing, whaling vessels were free to extend their hunting range, which in turn resulted in longer voyages. In the decades leading up to the American Revolution, the success of vessels hunting throughout the Atlantic Ocean saw the fleet grow. That war, however, had disastrous effects on the fishery, and by its end, nearly every American whaling vessel had been captured or destroyed. When the war finally ceased, whalers recycled any available hulls and, in anticipation of resurgence in the industry, began constructing new ships that were larger and had greater cargo capacities than their predecessors.

This increase in vessel size also brought changes to the preferred rig with brigs and then ships replacing smaller schooners. The addition of these new rigs resulted in the need for larger crews to handle the numerous and massive sails and to operate additional whaleboats. The rounding of Cape Horn and location of rich whaling grounds in the Pacific Ocean brought many more ships into the whaling business. Tensions between the United States and Britain, however, eventually led to the War of 1812, which again crippled the industry for its duration. Although numerous whaleships were destroyed during this time, American naval success in the Pacific region allowed American whalers to dominate the new grounds when operations resumed. The success of voyages in the early decades of the nineteenth century, and reports of seemingly unlimited whale stocks, once again led to growth in the size of whaling vessels. From the 1820s to the 1840s, the average whaler was a capacious, full-rigged ship averaging 350 tons and measured roughly 33 m in length, 8 m in beam, and 4 m in depth (Leavitt 1973:13). Though by this time many of these ships were purpose-built for whaling, recycled hulls often offered the most desirable characteristics of stability and cargo capacity.

The other physical features characterizing American whaleships of this period were the specific industrial components they shipped. The many industrial tools with which these vessels were equipped were integral components of the workplace and changed little from 1750 to around 1850. The three main physical features that distinguished whaleships from other types of this period were the sturdy whaleboats that hung from the heavy wooden davits along both sides of the ship, the cutting platform and heavy tackle used in lifting blubber pieces, and the immense brick tryworks. Each of these industrial elements specially related to one of the five steps defined in the whaling process, and their presence onboard a nineteenth-century sailing vessel offered unmistakable evidence that it engaged in the whale fishery.

From an operational standpoint, pelagic whaling of this period was characterized by its labor structure and wage system, as well as the outfitting of voyages. In their basic form, early nineteenth-century whaleships were sailing vessels that operated in the same way as those used for other nautical enterprises. Since the many sails associated with the ship's rig required numerous hands to work them, all contemporary vessels had large complements of mariners onboard. Whaleships, however, required even larger crews than vessels of similar size due to the numerous whaleboats they carried. While the whaler was underway and hunting with lookouts at the mast tops, its crews performed the duties of average sailors; when whales were sighted, however, the sails were taken in and all hands committed to their industrial roles.

As was common on all vessels of this period, the labor force was structured in a hierarchical manner that offered opportunities for advancement for individuals keen to progress through the ranks. By the mid-eighteenth century, whaling was seen as an admirable profession, and many young sailors followed in the wake of family members to pursue a life at sea. Although this tradition was still alive onboard most Nantucket whaleships of the early 1800s, as New Bedford grew in prominence and other opportunities presented themselves on the mainland, many of the more diligent young men were lured away from a career in whaling. Thus, as the nineteenth century progressed, owners and agents resorted to filling crews with inexperienced sailors who lacked the discipline or desire to engage in more than a single voyage.

The division of labor in this hierarchical system was physically manifested in the space provided for accommodation of the different positions of the crew at either end of the ship. The captain and officers were provided with relatively commodious, clean, and well-ventilated spaces at the stern. Forward of that was a slightly less comfortable area known as steerage, which was equipped with simple bunks for the boatsteerers, skilled craftworkers, and the steward. Located in the extreme forward end of the ship average sailors lived in the forecastle, which was characterized by numerous tiny bunks where up to 20 individuals slept and considered among the ship's most miserable places. The physical division onboard whaleships kept the officers from interacting with average mariners, which was strictly forbidden, and served to reinforce the rigid power structure.

Unless engaged in whaling activities, working periods were prescribed by the watch. Overseen by the officers, this system required every member of the crew to participate in an equal amount of work each day. Duties included maintaining lookouts, steering the ship, cleaning the deck, and any other simple chore. Crews not on watch engaged in recreational activities such as reading, writing, scrimshander, and fishing, which occurred in designated areas of the ship. When whales were sighted, the watch was suspended, and the whalers split into previously designated whaleboat crews. If one or more of the boats was successful in the hunt, returning to the ship meant that foremast hands carried out cutting in or trying out operations, all of which were supervised by the officers. After the blubber was processed, the oil was transferred into wooden casks and stowed in the hold. Before whaling operations were complete, the crew scrubbed the ship and prepared all necessary equipment for the next round of activity.

Status within the shipboard hierarchy was also distinguished at mealtimes through the same three-tiered structure that dictated accommodations. The dining experience of the officers was more refined than those of the crew, with their meals taken at a large table in the aft cabin and served on porcelain and silver place settings. Although the fare was often the same for all, officers were generally provided with a range of condiments and spices that greatly increased palatability. On some ships, when the officers finished their meals and vacated the dining cabin, the occupants of steerage dined at the table as well. In other cases, they received the leftovers of the officer's meals but were forced to eat using plain utensils in steerage. Average hands were provided meals generally considered less than appetizing and served in wooden bowls or tin pans. Unless they owned personal forks or spoons, a jackknife was the only utensil used for dining. Depending on the weather, their meals were taken on the ship's foredeck or in the grimy forecastle.

The financial compensation system was another organizational characteristic that distinguished whaling from many other nautical enterprises and helped to shape the workplace. Unlike merchant sailors of the time, American whalers were not paid a monthly wage; instead, they operated under the lay system that offered predetermined shares in profits. Used in many New England fisheries for centuries, it was an effective management strategy and motivator. The lays offered were relative to experience and position within the hierarchical labor structure. Although this system was designed to stimulate crews to work harder, any number of factors affected the profitability of a whaling cruise and the debts that crew members accrued. Thus, many average mariners were often lucky to receive any compensation at all for their efforts by the end of the voyage.

Aside from the labor and wage structure, the other critical organizational characteristic of the whale fishery involved supplying the ship. Although many of the contemporary commercial ships required large amounts of provisions since they engaged in long voyages across vast stretches of sea, none compared to the sheer magnitude of materials needed for a whaling voyage. With the installation of the tryworks, these ships undertook longer voyages to more distant grounds. Since the areas in which whalers hunted were generally remote and reprovisioning points were few, it was necessary for whaleships to carry large quantities of fresh and preserved food, fresh water, and extra industrial equipment to sustain extended sea time. To keep track of the various goods brought onboard, agents used detailed outfitting books. Since space was always a concern on whaleships, a clever system for storing the material shipped onboard was also devised, which involved packing supplies into casks that would later be used for oil storage. The casks were then stowed with detailed notes made regarding the contents and location of each. When needed, the materials stored in any given cask could be quickly accessed thanks to this innovative system.

To ensure productivity and maintain order in the cramped environment, pelagic whaling voyages were highly organized operations that involved strict adherence to the hierarchy of command and the structure of daily activities. As with the methods employed in the whaling process, the division of labor onboard these ships developed in the formative period of the mid- to late 1700s. Although the growth in ship size and changes in rig configuration resulted in the need for additional mariners, the basic structure of a whaleship crew and their responsibilities were defined by duties performed. Thus, the workplaces that existed onboard whaleships of the golden age were highly structured environments in which hard work and industrial success directly benefited not only the investors but every member of the crew.

"And So Ends This Day"

The wooden ships employed in American pelagic whaling in the early to mid-nineteenth century were a unique type of sailing vessel. Although in basic design and construction they were similar to those used in other mercantile activities of the period, whaleships were in fact complex floating industrial sites that acted as "mother ships" for several smaller hunting craft, provided a platform for industrial scale production, and served as a meager home for officers and crews for the duration of the multiyear voyages. Since most of the operations were conducted within the close confines of those relatively small vessels, a highly regimented and hierarchical system for management was employed. That system operated under three sequential stages that included both physical and organizational components designed to ensure maximum efficiency in all operational phases. Central to the pelagic whaling system were interrelated industrial processes, which developed over the middle decades of the eighteenth century and required strict adherence to prescribed methods and highly specialized tools. The effectiveness of this system allowed American whalers to hunt throughout the world's oceans with great success, the benefits of which had a major financial impact on the development of the United States and helped to establish American prominence in the Pacific Ocean.

This book illustrates the usefulness of integrating data derived from archival, archaeological, and material culture research for understanding the industrial nature of American pelagic whaling in the first half of the nineteenth century. Other fruitful avenues remain for studying nineteenth-century pelagic whaling. Since several countries also engaged in the fishery during this period and most adopted the tools and methods used in seaborne operations developed by American whalers, comparative studies of other fisheries with those of the United States could illicit interesting information about knowledge transfer, technological diffusion, and cultural adaptation. Of particular interest would be British whaling operations in the Pacific region, as the two nations were historically known to have influenced one another in many ways. Such comparative studies could be highly beneficial to understanding the similarities and differences between the two fisheries, and material culture studies of the known wrecks of British whaleships of contemporary vintage in the NWHI could provide an excellent opportunity to see how they manifest archaeologically.

Another potential area of research that could prove interesting is the application of themes generally used in industrial archaeology practice to shipwreck sites associated with other types of marine resource extraction. Although whaling was distinctive from other seaborne trades since it incorporated oil production facilities onboard the ships, trades such as fishing, pearling, and oyster harvesting might also be considered industrial in nature. Thus, the application of industrial research themes could help to better understand aspects such as organizational structure, conditions experienced by crews, methods and technologies employed, scopes of operation, and effects of environmental conditions on outcomes.

Appendix

Articles for a Whaling Voyage

Provisions and Cabin Stores.
Flour Baked
Flour packed
Mess Beef
Prime do.[a]
Mess Pork
Prime do.
Molasses
Sugar
Butter
Cheese
Rice
Coffee
Vinegar
Old Cider
Dried Apples
Codfish
Tongues and Sounds
Mackerel
Pickles
Raisens
Pork Hams
Smoked Beef
Kiln-dried Meal
Beans
Peas
Corn
Potatoes
Onions
Lamp Oil
Sperm Candles
Hard Soap
Oil Soap
Chocolate
Souchong Tea
Hyson do.
Loaf Sugar
Sweet Corn
Mustard
Mustard Seed
Black Pepper
Cayenne do.
Ginger
Allspice
Nutmegs
Cloves
Cinnamon
Selaratus
Sweet Oil
Pepper Sauce
Table Salt
Essence Spruce
Hops
Lemon Syrup
Sage
Summer Savory
Coarse Salt

FOR MEDICAL PURPOSES
N.E. Rum
H. Gin
Brandy
Port Wine

Miscellaneous.
Common tar
Coal tar
Rosin
Oars
Spruce Poles
Iron Poles
Saw dust
Sand
Old Lead
Sheet Lead
Charcoal
Sea Coal
Old Junk
Smith's Bellows
Spare Brick
Lime
Clay

Pine Heading
" Boards
" Plank
Oak do.
Spare Staves
White Oak Butts
Boats
Boat Boards
" Keels
Boat Knees
" Gunwales
" Stems
" Stern Posts
" Timbers
Fluke Chains
Fin do.
Iron Hoops
Grate Iron
Try pots
Deck Pots
Camboose and Apparatus
Cannon
Hose
Hose Tub
Line Tub
Boat Kegs
Harness Cask
Scuttle Butt
Chronometer
Barometer
Thermometer
Charts
Medicine Chest
Epsom Salts
Spare Spars

Compasses Large Boat
Small[b]

Hardware, & c.
Pitch Pot
Sauce Pans
Fry Pans
Grid Irons
Tea Kettles
Ladles
Tormentors
Skimmers
Shovels and Tongs
Scrubbing Brushes
Floor Brushes
Dust do.
Cooks Bellows
Chopping Knife
Chopping Tray
Dust Pan
Lamp Wick
Brimstone
Black Lead
Fire Steels
Flints
Bristol Brick
Rotten Stone
Glue
Coffee Mills
Britannia Ladles
do. Tureens
do. Teapots
do. Tumblers
do. Table Spoons
do. Tea do.
Iron Tea do.
do. Tea do.
do. Wash Basin
Coarse and Fine Seives [*sic*]
Cork Screws
Carver and Fork
Table Steel
Butchers Steels
Knives and Forks
Cabin Bell
Table do.
Cabin Lamps
Jacket do.
Socket do.
Brass Candlesticks
Mortar and Pestle
Time glasses
Spy Glass
Bread Tray
Looking Glass
Globe Lantern
Binnacle Lantern
do. Lamps
Hand Cuffs
Marlin Spike
Shot
Boat Corks
Palm Irons
Deep Sea Leads
Hand do.
Deep Sea Lines
Hand do.
Fish do.
Cod do.
Log do.
Fish Hooks
Houseline and Marline
Sewing Twine
Copper Pump Tacks
Iron do. do.
Iron Brads
Scupper Nails
Clout do.
Brass and Iron Screws
Scrapers
Steelyards
Butt Cock
Beer do.
Molasses Gates
Shovels
Hoes
Birch Brooms
Corn do.
Tar Brushes
Paint do.

Marking Brushes
Rigging Leather
Pump do.
Brad Awls
Sewing Awls
Table Covers
Shoe Thread
Log Book
Log Slates
Common Slates
Slate Pencils
Writing Paper
Ink
Ink Stand
Ink Powder
Quills
Signals
Jacks
Pendants
Ensigns
English Mincing Knives
Blubber Forks
Muskets

Sail Needles
Large Marline
Small do.
Large Bolt Rope
Middle do. do.
Small do. do.
Head Rope
Store do.
Tabline
Flat Seam

Nails
Clinch
Timber
Gunwhale
Lap
Foot
Ceiling
Wrought 20d, 10d, 8d, 6d.
Cut 20d, 10d, 8d, 6d.
Old Spikes

Cooper's Tools
Long Jointer
Jointer Irons
Truss Hoops
Axes
Adzes
Large Crose
Small do.
Crose Irons
Hollowing Plane
Heading do.
Bilge do.
Smooth do.
Levelling do.
Stock Howell
Draw Knives
Crooked do.
Champering [*sic*] Knives
Hampering do.
Bilge do.
Inshave [*sic*]
Dub Howell
Flagging Irons
Beek [*sic*] Irons
Marking Irons
Bung Borers
Tap do.
Compasses
Bitt Stock
Bitts
Vises
Guaging [*sic*] Rod
Red Cedar
Bungs
Chalk
Hammers
Drivers
Punches
Rivet Setts
Cold Chisels

Rivets
2d.
3d.
4d.
5d.
6d.

Iron Hoops
2d.
3d.
4d.
5d.
6d.

Carpenter's Tools
Long Jointers
Short do.
Jack Plane
Smooth Plane
Hollowing Plane
Rounding do.
Nail Hammers
Boat do.
Pump do.
Coopering Hammers
Broad Axe
Deck do.
Narrow Axe
Deck Hatchet
Boat Hatchet
Carpenter's Adze
Large Pincers
Small do.
Hand Saws
Back do.
Compass Saws
Wood do.
Drawing Knives
Spoke Shaves

Nail Gimlets
Spike do.
Ruff do.
Iron Square
Steel do.
Steel Tongue do.
Bevel
Carpenter's Compasses
Bench Vise
Hand Vises
Thumb do.
Marking Irons
Caulking do.
Screw Augers
Screw Divers
Saw Sett
Brace and Bitts
Old Stones
Water do.
Rifle do.
Sand do.
Rifles
Carpenter Rule
Flat Files
Half Round do.
Round do.
Wood Rasps
Three-square Files
Socket Chisels
do. Gouges
Firmer Chisels
do. Gouges
Scarf Chisels
Sets Chisels 1-8 to 1 inch
do. Chisels 1-8 to 1 inch
Board Gage
Square Plane Irons
Chalk Line
Trowell [*sic*]
Ruffs and Clinches
Pad Locks
Boat Knives
Leaning do.
Bullet Moulds
Blacksmith's Anvil
Sand Paper
Grind Stones

Kegs Powder[b]
Musket
Pistol

Crockery
Shoal Plates
Soup do.
Small do.
Pudding Dishes
Oval do.
Gravy do.
Soup Tureen
Sugar Bowl
Butter Tubs
Castor
Castor Bottles
Salts
Tumblers
Wines
Decanters
Bowls
Covered Dishes
Pitchers
Platters
Mugs and Saucers
Cups and Saucers
Spare Deck Lights

Tin Ware
Signal Lanterns
Boat do.
Cooks do.
Ladles
Dippers
Savealls
Scoops
Tinder Boxes
Tin Cups
" Pots
" Pans
Bake Pans
Pie Plates
Graters
Tunnels
Tean Cannisters
Oil do.
Powder do.
Molasses Cup
Protection Box
Soup Tureen
Trumpets
Blow Horn
Measures
Pudding Bag
Coffee Pots
Tea Pot
Lamp Feeders
Blubber Room Lamps
Wash Basin
Pepper Boxes
Gudgeon Boxes

Flour Scoop[b]
Dust Pan
Spitoon

Copper Ware
Cooler
Dipper
Skimmers
Pumps
Ladles
Tunnels
Tunnel Noses
Savealls
Vent Pipe
Cooler Cock
Hose do.

do. Joints
Tub Screw
Brass Trumpet
Binnacle Bell

Craft, &c.
Whale Irons
Lances
Head Spades
Boat do.
Blubber Hooks
Boat do.
Boarding Knives
Stearing Braces
Grindstone Cranks
Pikes
Gafts
Line Hooks
Lilley Irons
Grains
Can Hooks
Chain Punches
Shackle Pins
Lance Hooks
Marking Irons
Crow Bars. Top Mall.
Tew Iron. Spike Tools
Rivet Tools
Blacksmith's Tools
do. Hammers

Cordage
Manilla Lines
Tarred do.
Manilla Cutting Falls
do. Cordage
Tarred do.
Wormline
Lance Line
Bolt Rope
Point Rope
Manilla Hawser
Tarred do.

Articles for Recruits
Tobacco
Cigars
Pipes
Powders
Muskets
Flints
Narrow Axes
Bleached Cotton
Unbleached do.
Blue do.
Furniture Prints
Fancy do.
Oil Soap
Bar do.
Lemon Syrup
English Green
Black Paint
White Lead
Linseed Oil

Spare Paints
White Lead
Black Paint
English Green
French Yellow
Verdigris
Prussian Blue
Vermillion
Crome Yellow
Venish Red
Japan
Linseed Oil
Spirits Turpentine
Copal Varnish
Spare Glass
Whiting
Bright Varnish

Blocks, &c.
Cutting Blocks
Burthen do.
Purchase do.
Guy do.
Spare do.
Cat do.
Shieves [*sic*]
Pins
Belaying Pins
Hanks
Pump Boxes
do. Breaks
Hand Spikes
Lance Poles
Spade do.

Slops
Pea Jackets
Blue Monkey Jackets
Drab do. do.
Blue Kersey Short do.
Drab do. do. do.
Satinett [*sic*] do. do.
Blue Thick Trousers
Drab Vermont do.
Satinett do.
Duck do.
Vests
Red Twilled Kersey Shirts
Striped do. do. do.
do. Cotton do.
Calico Full Bloom do.
do. Plain do.
Stockings
Mittins [*sic*]
Gurnsey [*sic*] Frocks
Duck do.
Cotton and Wool Drawers
Wool do.
Tarpaulin Hats

Blankets
Comforters
Braces
Cotton Handkerchiefs
do. Shawls
Thread
Scotch Caps
Sheaths and Belts
Sheath Knives
Jack do.
Palm Irons
Iron Spoons
Tin Pots
do. Pans
Shoes
Pumps

Spare Sails
Flying Jibb
Jibb
F.T.M. Staysail
Foresail
Fore Topsail
Fore T.G. Sail
Mainsail
Main Topsail
Main T.G. Sail

Main Spencer
Mizzen Topsail
Mizzen T.G. Sail
Spanker
T.G. Stud Sails
Lower do. do.

Spare Duck
Thick Duck
Heavy Ravens
Thin Duck

Source: New Bedford Whaling Museum Research Library and Archives, 1832.
[a] Single quotation mark and the abbreviation "do." were commonly used in outfitting books to indicate the repeat of a word directly above another on a list.
[b] These items were not preprinted; they are penciled in at some point during outfitting.

References Cited

Abernathy, William J., and James M. Utterback

1978 Patterns of Industrial Innovation. *Technology Review* 80(7):40–47.

Adams, James T.

1918 *A History of the Town of Southampton (East of Canoe Place).* Hampton Press, Long Island, New York.

Adams, Jonathan

2001 Ships and Boats as Archaeological Source Material. *World Archaeology* 32(3):292–310.

Aguilar, Alex

1986 A Review of Old Basque Whaling and Its Effect on the Right Whales (*Eubalaena glacialis*) of the North Atlantic. Special Issue, *Report International Whaling Commission* 10:191–199.

Alden, Dauril

1964 Yankee Sperm Whalers in Brazilian Waters, and the Decline of the Portuguese Whale Fishery (1773–1801). *Americas* 20(3):267–288.

Allen, Everett S.

1973 *Children of the Light: The Rise and Fall of New Bedford Whaling and the Death of the Arctic Fleet.* Little, Brown, Boston.

Allen, Glover M.

1916 *The Whalebone Whales of New England.* Boston Society of Natural History, Boston.

1928 Whales and Whaling in New England. *Scientific Monthly* 27(4): 340–343.

Ambrose, Edwin R.

2000 Early Whaling and the Development of Onboard Tryworks. *Dukes County Intelligencer* 42(2):87–90.

Amerson, A. Binion, Jr.

1971 The Natural History of French Frigate Shoals, Northwestern Hawaiian Islands. *Atoll Research Bulletin* 150.

Amerson, A. Binion, Jr., Roger C. Clapp, and William O. Wirtz II

1974 The Natural History of Pearl and Hermes Reef, Northwestern Hawaiian Islands. *Atoll Research Bulletin* 174.

Anderson, Ross

2004 Whaling and Colonial Trade on the Shipwreck *Cheviot* (1827–1854). *Bulletin of the Australian Institute for Maritime Archaeology* 28:1–10.

Anderson, Ross, and Madeleine McAllister

2012 *Koombana Bay Foreshore Maritime Archaeological Survey and Excavations 21–28 November 2011*. Department of Maritime Archaeology, Western Australian Maritime Museum, Fremantle, Australia.

Ansel, Willits Dyer

1983 *The Whaleboat, A Study of Design, Construction and Use from 1850 to 1970*. Mystic Seaport Museum, Mystic, Connecticut.

Appleby, John C.

2008 Conflict, Competition and Cooperation: The Rise and Fall of the Hull Whaling Trade during the Seventeenth Century. *Northern Mariner* 18(2):23–59.

Arlov, Thor B.

1993 Whaling and Sovereignty: The Role of Whaling in the Struggle for Supremacy over Svalbard (Spitsbergen). In *Whaling and History: Perspectives on the Evolution of the Industry*, edited by Jan E. Ringstad, Einar Wexelsen, and Bjørn L. Basberg, pp. 81–90. Kommadør Chr. Christiansens Hvalfangstmuseum, Sandefjord, Norway.

Arnold, J. A.

1912 *Yearbook of the United States Department of Agriculture 1911*. US Government Printing Office, Washington, DC.

Ashley, Clifford W.

1926 *The Yankee Whaler*. Dover Publications, New York.

Atkinson, Karen

1987 The Significance of the Rowley Shoals Wreck to the Study of Whaling in the South Seas. *Bulletin of the Australian Institute for Maritime Archaeology* 11(2):1–7.

Atlas and Daily Bee [Boston]

1859 Whalers. May 17:34.

Babits, Lawrence E., and Hans Van Tilburg (editors)

1998 *Maritime Archaeology: A Reader of Substantive and Theoretical Contributions*. Plenum Press, New York.

Badcock, Anna, and Brian Malaws

2004 Recording People and Processes at Large Industrial Structures. In *The Archaeology of Industrialization*, edited by David Barker and David Cranstone, pp. 269–290. Taylor & Francis Group, Leeds, UK.

Bailey, R. C.

1953 Shore Whaling Came First. *Long Island Forum* 16:83–85.

Baker, A. H.

1974 Chandlery and General Merchandise: The Outfitting and Provisioning of Mystic Whaleships in the Nineteenth Century. D. W. Blunt Research Library at Mystic Seaport, Mystic, Connecticut.

Baker, William A.

1973 *A Maritime History of Bath, Maine and the Kennebec River Region*. Vol. 2. Maritime Research Society of Bath, Bath, Maine.

1984 *The Lore of Sail*. Facts on File, New York.

Barber, Edwin A.

1907 *Salt Glazed Stoneware Germany, Flanders, England and the United States*. Doubleday, Page, New York.

1909 *The Pottery and Porcelain of the United States: A Historical Review of American Ceramic Art from the Earliest Times to the Present Day*. G. P. Putnam and Sons, New York.

Barker, Fen G. (editor)

1879 *A Complete Schedule of Vessels Built and Registered or Enrolled in the District of Bath, Maine, Commencing at 1781, Giving Rig, Name, Tonnage, Where Built, First Master, Registering or Enrolling Owner and Hailing Port*. Fen. G. Barker, Bath, Maine.

Barkham, Michael M.

1981 *Life on Board a 16th-Century Basque Whaler*. Parks Canada, Ottawa.

Barkham, Selma H.

1984 The Basque Whaling Establishments in Labrador 1536–1632—A Summary. *Arctic* 37(4):515–519.

Barthelmess, Klaus

2009 Basque Whaling in Pictures, 16th–18th century. *Itsas Memoria: Revista de Estudios Maritimos del Pais Vasco* 6:643–667.

Bartky, Ian R.

2000 *Selling the True Time: Nineteenth-Century Timekeeping in America*. Stanford University Press, California.

Basberg, Bjørn L.

1998 The Floating Factory: Dominant Design and Technological Development of Twentieth-Century Whaling Factory Ships. *Northern Mariner* 8(1):21–37.

2004 *The Shore Whaling Stations of South Georgia: A Study in Antarctic Industrial Archaeology*. Novus Forlag, Oslo, Norway.

Bass, George F.

1967 Cape Gelidonya: A Bronze Age Shipwreck. *Transactions of the American Philosophical Society* 57:8.

1983 A Plea for Historical Particularism in Nautical Archaeology. In *Shipwreck Anthropology*, edited by Richard A. Gould, pp. 91–104. University of New Mexico Press, Albuquerque.

Baxter, R. Scott

2002 Industrial and Domestic Landscapes of a California Oil Field. *Historical Archaeology* 36:18–27.

Beaton, Kendall

1955 Dr. Gesner's Kerosene: The Start of American Oil Refining. *Business History Review* 29(1):28–53.

Bellwood, Peter

1979 *Man's Conquest of the Pacific*. Oxford University Press, New York.

Biddlecombe, George

1925 *The Art of Rigging: Containing an Explanation of Terms and Phrases and the Progressive Method of Rigging Expressly Adapted for Sailing Ships*. Dover Publications, Mineola, New York.

Binford, Lewis R.

1962 Archaeology as Anthropology. *American Antiquity* 28(2):217–225.

Bingeman, John M., John P. Bethel, Peter Goodwin, and Arthur T. Mack

2000 Copper and Other Sheathing in the Royal Navy. *International Journal of Nautical Archaeology* 29(2):218–229.

Birkett, Mary E.

2000 Hawai'i in 1819: An Account by Camille de Roquefeuil. *Hawaiian Journal of History* 34:69–92.

Blackman, David J.

2000 Is Maritime Archaeology on Course? *American Journal of Archaeology* 104(3):591–596.

Blainey, Geoffrey

1966 *The Tyranny of Distance: How Distance Shaped Australia's History*. Sun Books, Melbourne, Australia.

Bockstoce, John R.

1984 From Davis Strait to Bering Strait: The Arrival of the Commercial Whaling Fleet in North America's Western Arctic. *Arctic* 37(4):528–532.

2006 Nineteenth Century Shipping Losses in the Northern Bering Sea, Chukchi Sea and Beaufort Sea. *Northern Mariner* 16(2):53–68.

Boggs, S. Whittemore

1938 American Contributions to the Geographical Knowledge of the Central Pacific. *Geographical Review* 28(2):177–192.

Bolster, W. Jeffrey

1990 "To Feel Like a Man": Black Seamen in the Northern States, 1800–1860. *Journal of American History* 76(4):1173–1199.

Boston Daily Journal

1859 Ship News. May 14:28.

Botwick, Bradford, and Debra A. McClane

2005 Landscapes of Resistance: A View of the Nineteenth Century Chesapeake Bay Oyster Fishery. *Historical Archaeology* 39(3):94–112.

Bowditch, Nathaniel

1817 *New American Practical Navigator*. Cushing & Appleton, Salem, Massachusetts.

Braat, J.

1984 Dutch Activities in the North and the Arctic during the Sixteenth and Seventeenth Centuries. *Arctic* 37(4):473–480.

Bradley, Charles S.

1993 Ship's Fittings and Rigging Components Recovered from the 1983

Underwater Archaeological Excavation at Red Bay, Labrador. *Research Bulletin* 300. Parks Canada.

Bradley, Harold W.

1942 *The American Frontier in Hawaii: The Pioneers 1789–1843.* Peter Smith, Gloucester, Massachusetts.

1943 Hawaii and the American Penetration of the Northeast Pacific, 1800–1845. *Pacific Historical Review* 12(3):277–286.

Braginton-Smith, John, and Duncan Oliver

2008 *Cape Cod Shore Whaling: America's First Whalemen.* History Press, Charleston, South Carolina.

Bratten, John R.

2002 *The Gondola* Philadelphia *and the Battle of Lake Champlain.* Texas A&M Press, College Station.

Brimley, Herbert H.

1895 Whale Fishing on the Coast of North Carolina. *American Angler* 25(3):67–75.

Bronson, George W.

1855 *Glimpses of the Whaleman's "Cabin."* Damrell and Moore, Printers, Boston.

Brooke, George M., Jr. (editor)

1986 *John M. Brooke's Pacific Cruise and Japanese Adventure, 1858–1860.* University of Hawaii Press, Honolulu.

Brooke, J. M.

1859 Lisansky's Island. G9237 H5 185 H38. Library of Congress Geography and Map Division.

Brooks, N. C.

1860 Islands and Reefs W.N.W. of the Sandwich Islands, Pacific. *Nautical Magazine and Naval Chronicle for 1860: A Journal of Papers on Subjects Connected with Maritime Affairs*, 499–504.

Brown, C. F.

1835 Journal, October 10, 1831–February 23, 1835. Nicholson Whaling Log Collection Number 540. Providence Public Library, Providence, Rhode Island.

Brown, James T.

1883 *The Whale Fishery and Its Appliances.* US Government Printing Office, Washington, DC.

1887 The Whalemen, Vessels, Apparatus, and the Methods of the Fishery. In *The Fisheries and Fishery Industries of the United States*, edited by George B. Goode, vol. 2, pp. 218–293. US Government Printing Office, Washington, DC.

Browne, John R.

1850 *Etchings of a Whaling Cruise, with Notes of a Sojourn to the Island of Zanzibar, to Which Is Appended a Brief History of the Whale Fishery, Its Past and Present Condition.* Harper & Brothers, New York.

Brunal-Perry, O., M. G. Driver, and T. L. Carrell

2009 Early European Exploration and the Spanish Period in the Marianas

1521–1898. In *Maritime History and Archaeology of the Commonwealth of the Northern Mariana Islands*, edited by Toni Carrell, pp. 61–94. Publisher, Saipan, Northern Mariana Islands.

Brundage, B. C.

1948 The Early American Whale Fishery. *Historian* 11(1):54–72.

Buck, Peter H.

1953 Explorers of the Pacific. *Bishop Museum Special Publication* 43.

Bullen, Frank T.

1898 *The Cruise of the Cachalot: Round the World After Sperm Whales.* Grossett & Dunlap, New York.

Burns, Jason M.

2003 *The Life and Times of a Merchant Sailor: The Archaeology and History of the Norwegian Ship* Catharine. Kluwer Academic/Plenum, New York.

Calkin, Milo

1953 Milo Calkin's Journal 1833–1842. Hawaii and Pacific Collections. University of Hawaii at Manoa, Honolulu.

Campbell, George Frederick

1974 *China Tea Clippers.* Donald McKay, New York.

Canadian Parks Service

1992 *The Wreck of the* Auguste. Parks Canada, Ottawa.

Carter, John, and T. J. Kenchington

1985 The Terence Bay Wreck: Survey and Excavation of a Mid-18th Century Fishing Schooner. In *Proceedings of the Sixteenth Conference on Underwater Archaeology*, edited by P. F. Johnston, pp. 13–26. Ann Arbor, Michigan.

Casella, Eleanor C.

2001 Every Procurable Object: A Functional Analysis of the Ross Factory Archaeology Collection. *Australasian Historical Archaeology* 19:25–38.

2006 Transplanted Technologies and Rural Relics: Australian Industrial Archaeology and Questions that Count. *Australasian Historical Archaeology* 24:65–75.

Casserley, Tane R.

1998 *A Maritime History of the Northwestern Hawaiian Islands from Laysan to Kure.* Unpublished report on file, Marine Option Program, University of Hawaii at Manoa, Manoa.

Cederlund, Carl O.

2006 Vasa I. *The Archaeology of a Swedish Warship of 1628.* Edited by Fred Hocker. National Maritime Museum of Sweden, Stockholm.

Chamberlain, S.

1988 The Hobart Whaling Industry 1830–1900. Unpublished PhD dissertation, La Trobe University, Melbourne, Australia.

Chan, H. L.

2020 Economic Impacts of Papahanaumokuakea Marine National Monument Expansion on the Hawaii Longline Fishery. *Marine Policy* 115:1–13.

Chandler, P. W., and S. H. Phillips

1848 The Ship *Holder Borden. Law Reporter* 10:194–201.

Chapelle, Howard I.

1967 *The Search for Speed under Sail, 1700–1855.* W. W. Norton, New York.

1973 *The American Fishing Schooners 1825–1935.* W. W. Norton, New York.

Chappel, Thomas

1824 *An Account of the Loss of the* Essex. Religious Tract Society, London.

Chase, Owen

1821 *Narrative of the Most Extraordinary and Distressing Shipwreck of the Whale-ship* Essex, *of Nantucket.* W. B. Gilley, New York.

Chatterton, Edward K.

1925 *Whalers and Whaling: The Story of the Whaling Ships up to the Present Day.* T. Fisher Unwin, London.

Chatwin, Dale

1996 "A Trade So Uncontrollably Uncertain": A Study of the English Southern Whale Fishery from 1815 to 1860. Unpublished master's thesis, Australian National University, Canberra.

Cheever, H. T.

1850 *The Whale and Its Captors; or, The Whalemen's Adventures, and The Whale's Biography, as Gathered on the Homeward Cruise of the "Commodore Preble."* Harper and Brothers, New York.

Cherfas, Jeremy

1988 *The Hunting of the Whale: A Tragedy That Must End.* Bodley Head, London.

Childs, S. Terry, and David Kirkpatrick

1993 Indigenous African Metallurgy: Nature and Culture. *Annual Review of Anthropology* 22:317–337.

Chochorowski, Jan

1991 Archaeology in the Investigation of the History of Human Activity in the Region of Spitsbergen. *Polish Polar Research* 12(3):391–406.

Church, Albert C.

1938 *Whale Ships and Whaling.* W. W. Norton, New York.

Clapham, Philip J., and C. Scott Baker

2002 Modern Whaling. In *Encyclopedia of Marine Mammals*, edited by William F. Perrin, Bernd Würsig, and J. G. M. Thewissen, pp. 1328–1332. Academic Press, New York.

Clapp, Roger B.

1972 The Natural History of Gardner Pinnacles, Northwestern Hawaiian Islands. *Atoll Research Bulletin* 163.

Clapp, Roger B., E. Kridler, and R. R. Fleet

1977 The Natural History of Nihoa Island, Northwestern Hawaiian Islands. *Atoll Research Bulletin* 207:1–147.

Clapp, Roger B., and William O. Wirtz II

1975 The Natural History of Lisianski Island, Northwestern Hawaiian Islands. *Atoll Research Bulletin* 186.

Clark, A. H.

1887 History and Methods of the Fisheries. In *The Fisheries and Fishery*

Industries of the United States, edited by George B. Goode, vol. 2, pp. 3–218. US Government Printing Office, Washington, DC.

Clark, C. M.

1987 Trouble at T'Mill: Industrial Archaeology in the 1980s. *Antiquity* 61:169–179.

Clement, Russell

1980 From Cook to the 1840 Constitution: The Name Change from Sandwich to Hawaiian Islands. *Journal of Hawaiian History* 14:50–57.

Coles, Harry L.

1965 *The War of 1812*. University of Chicago Press, Chicago, Illinois.

Connah, Graham

1994 *The Archaeology of Australia's History*. Cambridge University Press, United Kingdom.

Conway, William M.

1902 *Early Dutch and English Voyages to Spitsbergen in the 17th Century*. Hakluyt Society, London.

Cooney, Gabriel

2003 Introduction: Seeing Land from the Sea. *World Archaeology* 35(3):323–328.

Cossons, Neil

2005 New Directions in Industrial Archaeology. In *Industrial Archaeology: Future Directions*, edited by Eleanor C. Casella and James Symonds, pp. ix–x. Springer, New York.

Costell, George

1856 *A Treatise on Ships' Anchors*. John Weale, London.

Coulter, John W.

1964 Great Britain in Hawaii: The Captain Cook Monument. *Geographical Journal* 130(20):256–261.

Couthouy, Joseph P.

1844 Remarks upon Coral Formations in the Pacific; With Suggestions as to the Causes of their Absence in the Same Parallels of Latitude on the Coast of South America. *Boston Journal of Natural History* 4:67–106.

Cranstone, David

2001 Industrial Archaeology—Manufacturing a New Society. In *The Historical Archaeology of Britain, c. 1540–1900*, edited by Richard Newman, David Cranstone, and Christine Howard-Davis, pp. 183–210. Routledge, London.

2004 The Archaeology of Industrialization—New Directions. In *The Archaeology of Industrialization*, edited by David Barker and David Cranstone, pp. 313–320. Maney, Leeds.

Credlund, Arthur G.

1989 Moby Dick, Hull and East Yorkshire. *Great Circle* 11(1):44–54.

Creighton, Margaret S.

1990 Fraternity in the American Forecastle 1830–1870. *New England Quarterly* 63(4):531–557.

Crothers, William L.
1997 *The American Built Clipper Ship: 1850–1856.* International Marine Press, Camden, Maine.

Crumlin-Pedersen, Ole, and Olaf Olsen
1978 *Five Viking Ships from Roskilde Fjord.* Roskilde National Museum, Roskilde, Denmark.

Daily Mercury [New Bedford, Massachusetts]
1845 Visit of the Brig *Delaware* to Pell's Island. March 21:14.
1867 Marine Journal. June 25:36.

Daily National [San Francisco, California]
1859 Loss of an American Whaleship. May 2:2.

Dakin, William J.
1934 *Whalemen Adventures: The Story of Whaling in Australian Waters and Other Southern Seas Related Thereto, from the Days of Sails to the Modern Times.* Angus & Robertson, Sydney, Australia.

Dana, Richard H., Jr.
1842 *Two Years before the Mast: A Personal Narrative of Life at Sea.* Harper & Brothers, New York.

Dana, Thomas F.
1971 On the Coral Reefs of the World's Most Northern Atoll (Kure: Hawaiian Archipelago). *Pacific Science* 25:80–87.

Daum, Arnold R.
1959 Petroleum in Search of an Industry. *Pennsylvania History* 26(1):21–34.

Davis, Charles G.
1918 *The Building of a Wooden Ship.* United States Shipping Board Emergency Fleet Corporation, Philadelphia.

Davis, Lance E., and Robert E. Gallman
1994 American Whaling, 1820–1900: Dominance and Decline. In *Whaling and History: Perspectives on the Evolution of the Industry,* edited by Jan E. Ringstad, Einar Wexelsen, and Bjørn L. Basberg, pp. 55–65. Kommandør Chr. Christensens Hvalfangstmuseum, Sandefjorde, Norway.

Davis, Lance E., Robert E. Gallman, and Karin Gleiter
1997 *In Pursuit of Leviathan: Technology, Institutions, Productivity, and Profits in American Whaling 1816–1906.* University of Chicago Press, Chicago.

Davis, Lance E., Robert E. Gallman, and Teresa D. Hutchins
1987 Productivity in American Whaling: The New Bedford Fleet in the Nineteenth Century. *National Bureau of Economic Research Working Paper,* no. 2477.
1988 The Decline of U.S. Whaling: Was the Stock Running Out? *Business History Review* 62(4):569–595.

Davis, William M.
1874 *Nimrod of the Sea; or, The American Whaleman.* Harper & Brothers, New York.

Davoll, Edward S.
1981 *The Captain's Specific Orders on the Commencement of a Whale Voyage to*

his Officers and Crew. Old Dartmouth Historical Society, New Bedford, Massachusetts.

Daws, Gavan

1968 *Shoal of Time: A History of the Hawaiian Peoples*. University of Hawaii Press, Honolulu.

Day, Arthur G.

1966 *Explorers of the Pacific*. Duell, Sloan and Pearce, New York.

de Zulueta, Julian

2000 The Basque Whalers: The Source of Their Success. *Mariner's Mirror* 86(3):261–271.

Decker, Robert O.

1974 *Whaling Industry of New London*. Liberty Cap Books, York, Pennsylvania.

Dégros, Maxime

1940 La Grande Pêche Basque des Origines à la Fin du XVIIIe Siècle. *Bulletin de la Société des Science, des Arts et Lettres de Bayonne* 35:148–179.

Delano, Reuben

1846 *Wanderings and Adventures of Reuben Delano, Being a Narrative of Twelve Years Life on a Whaleship!* Thomas Drew, Jr., Worcester, Massachusetts.

Delgado, James P.

1980 No Longer a Buoyant Ship: Unearthing the Gold Rush Storeship *Niantic*. *California History* 58(4):316–325.

2006 Gold Rush Entrepôt: The Maritime Archeology of the Rise of the Port of San Francisco. Unpublished PhD dissertation, Simon Fraser University, Burnaby, British Columbia.

2009 *Gold Rush Port: The Maritime Archeology of San Francisco's Waterfront*. University of California Press, Berkeley.

Dellino-Musgrave, Virginia E.

2006 *Maritime Archaeology and Social Relations: British Action in the Southern Hemisphere*. Springer Press, New York.

Department of Land and Natural Resources (DLNR)

2005 Title 13, Chapter 60.5: Northwestern Hawaiian Islands Marine Refuge. State of Hawaii.

Desmond, Charles

1919 *Wooden Shipbuilding*. Rudder Publishing, New York.

Dickerman, Marion

1949 *The Story of the Last of the Old Whalers:* Charles W. Morgan. Marine Museum of the Marine Historical Association, Mystic, Connecticut.

Dickson, Rod

2011 *Hezekiah Pinkham: His Jurnell.* Published by the author, Perth, Australia.

Doane, B.

1987 *Following the Sea: A Young Sailor's Account of the Seafaring Life in the Mid-1800s*. Nimbus Publishing and the Nova Scotia Museum, Halifax.

Dodd, A.

2013 *Norwegian Whaler's Base (1926–1932), Rakiura/Stewart Island: Marine*

Archaeological Survey of the "The Base" Prices Inlet for Project Njord 2013. New Zealand Heritage Places Trust, Wellington.

Dodge, Ernest S.

1965 *New England and the South Seas.* Harvard University Press, Cambridge, Massachusetts.

1971 *Beyond the Capes: Pacific Exploration from Captain Cook to the Challenger 1776–1877.* Victor Gollancz, London.

1976 *Islands and Empires: Western Impact on the Pacific and East Asia.* Burns and MacEachern. Don Mills, Ontario.

Dodge, George A.

1882 *A Whaling Voyage in the Pacific Ocean.* Merrill & Mackintire, Salem, Massachusetts.

Dolin, Eric J.

2007 *Leviathan: The History of Whaling in America.* W. W. Norton, New York.

Dorsett, E. L.

1954 Hawaiian Whaling Days. *American Neptune* 14(1):42–46.

Dow, George F.

1925 *Whale Ships and Whaling: A Pictorial History.* Dover Publications, Mineola, New York.

Draper, Charla L.

2001 *Cooking on Nineteenth-Century Whaleships.* Blue Earth Books, Mankato, Minnesota.

Dublin Penny Journal

1836 The Whale Fishery—Captain Ross. 4(200):347–348.

Dulles, Foster R.

1933 *Lowered Boats: A Chronicle of American Whaling.* Harcourt, Brace, New York.

du Monceau, Henri L. D.

1782 *Traité Général des Pesches, et Histoire des Poissons Qu'elles Fournissent, Tant pour la Subsistance des Hommes que pour Plusieurs autres Usages qui ont Rapport aux Arts et au Commerce.* Vol. 4. Saillant & Nyon, Paris.

Dupont, Ralph P.

1954 The *Holder Borden. New England Quarterly* 27(3):355–365.

Dyer, Michael P.

1999 The Historical Evolution of the Cutting-In Pattern, 1798–1967. *American Neptune* 59:137–148.

East, W. G.

1931 The Port of Kingston-upon-Hull during the Industrial Revolution. *Economica* 32:190–212.

Edwards, Everett J., and Jeanette E. Rattray

1932 *"Whale Off!" The Story of American Shore Whaling.* Coward-McCann, New York.

Ellis, Richard

1991 *Men and Whales.* Alfred A. Knopf, New York.

English Heritage

2006 *Understanding Historic Buildings: A Guide to Good Recording Practice.* English Heritage, Swindon, United Kingdom.

Erskine, Nigel

1997a Construction Analysis of the *Day Dawn*. *Bulletin of the Australian Institute for Maritime Archaeology* 21:111–116.

1997b Survivors from the Wreck of the *Day Dawn*. *Bulletin of the Australian Institute for Maritime Archaeology* 21:117–118.

Fairburn, William A.

1945 *Merchant Sail.* Fairburn Marine Educational Foundation, Center Lovell, Maine.

Fautin, Daphne, Penelope Dalton, Lewis S. Incze, Jo-Ann C. Leong, C. Pautzke, Andrew Rosenberg, Paul Sandifer et al.

2010 An Overview of Marine Biodiversity in United States Waters. *PLOS ONE* 5(8):47.

Finckenor, George A.

1975 *Whales and Whaling: Port of Sag Harbor.* William Ewers Printers, Sag Harbor, New York.

Finney, S. S.

2010 Toward an Understanding of Nineteenth Century Whaling Practices: A Risk Sensitivity Model for Whaling Agents and Masters. Unpublished PhD dissertation, University of Hawaii, Honolulu.

Finney, S. S., and M. W. Graves

2002 *Site Identification and Documentation of a Civil War Shipwreck Thought to be Sunk by the C.S.S.* Shenandoah *in April 1965.* American Battlefield Protection Program, National Park Service, Washington, DC.

Fischer, Steven R.

2002 *A History of the Pacific Islands.* Palgrave Macmillan, New York.

Flatman, Joe

2003 Cultural Biographies, Cognitive Landscapes and Dirty Old Bits of Boat: "Theory" in Maritime Archaeology. *International Journal of Nautical Archaeology* 32(2):143–157.

Fonda, Douglass C., Jr.

1969 Eighteenth Century Nantucket Whaling: As Compiled the Original Logs and Journals of the Nantucket Athenaeum and the Nantucket Whaling Museum. Published by author, Nantucket, Massachusetts.

Ford, Ben (editor)

2011 *The Archaeology of Maritime Landscapes.* Springer, New York.

Ford, Ben, Amy Borgens, William Bryant, Dawn Marshall, Peter Hitchcock, Cesar Arias, and Donny Hamilton

2008 Archaeological Excavation of the Mardi Gras Shipwreck (16GM01), Gulf of Mexico Continental Slope. Report on file, US Department of the Interior, Minerals Management Service, Gulf of Mexico OCS Region, New Orleans, Louisiana.

Forman, Henry C.

1966 *Early Nantucket and Its Whale Houses.* Mill Hill Press, Nantucket, Massachusetts.

Fox, Georgia L.

2006 *Final Report to the National Oceanic and Atmospheric Administration of the Conservation and Stabilization of the Northwestern Hawaiian Islands Marine Sanctuary Artifacts by the Heritage Resource Conservation Laboratory at California State University, Chico, Department of Anthropology.* National Oceanic and Atmospheric Administration and Papahanaumokuakea Marine National Monument.

2010 *Final Report to the National Oceanic and Atmospheric Administration (NOAA): The Conservation of Two Ships' Bells and Sounding Lead.* National Oceanic and Atmospheric Administration and Papahanaumokuakea Marine National Monument.

2012 *Final Report to the National Oceanic and Atmospheric Administration (NOAA): The Conservation of the* Two Brother's *Artifacts.* National Oceanic and Atmospheric Administration and Papahanaumokuakea Marine National Monument.

Francis, Daniel

1991 *The Great Chase: A History of World Whaling.* Penguin Books, Toronto.

Franck, S. M.

1986 The Legacy of Stranded Whales, Part Two: Tryworks at the Kendall Whaling Museum. *Whalewatcher* 20(3):5–9.

Franklin, Benjamin

1786 Maritime Observations and a Chart of the Gulph Stream. *Transactions of the American Philosophical Society* 2(38):294–329.

Franklin, Marianna

2005 Blood and Water; The Archaeological Excavation and Historical Analysis of the Wreck to the *Industry*, a North-American Transport Sloop Chartered by the British Army at the End of the Seven Years' War: British Colonial Navigation and Trade to Supply Spanish Florida in the Eighteenth Century. Unpublished PhD dissertation, Department of Anthropology, Texas A&M University, College Station.

Franzen, John G.

1992 Northern Michigan Logging Camps: Material Culture and Worker Adaptation on the Industrial Frontier. *Historical Archaeology* 26:74–98.

Freeman, Otis W. (editor)

1951 *Geography of the Pacific.* John Wiley and Sons, New York.

Freidel, Frank

1943 A Whaler in Pacific Ports, 1841–1842. *Pacific Historical Review* 12(4):380–390.

Friedlander, Alan M., and Edward E. DeMartini

2002 Contrasts in Density, Size, and Biomass of Reef Fishes between the Northwestern and the Main Hawaiian Islands: The Effects of Fishing Down Apex Predators. *Marine Ecology Progress Series* 230:253–264.

Friend [Honolulu, Hawaii]

1844 Wreck of the Whaleship *Holder Borden*. October 9, 2(10):93.

1846 Loss of the American Whaleship *Konohassett*. August 15, 4(16):124.

Friis, Herman R.

1967 The Pacific Basin: A History of Its Geographical Exploration. *American Geographical Society Special Publication* 38.

Gale Huntington Research Library, Martha's Vineyard Museum (MVM)

1765 Outfit of a Whaling Vessel in 1765. RU207, Box 5, Number 6.

Galtsoff, Paul

1931 The U.S.S. *Whippoorwill* Expedition to Pearl and Hermes Reef. *Mid-Pacific Magazine* 41(1):49–56.

Gardner, E.

1823 February 11, 1822, Sea Account. Nantucket Historical Association Research Library,

Gesner, Abraham

1861 *A Practical Treatise on Coal, Petroleum, and Other Distilled Oils*. Bailliere Brothers, New York.

Giambarba, Paul

1967 *Whales, Whaling and Whalecraft*. Scrimshaw Publishing, Centerville, Massachusetts.

Gibbins, David

1990 Analytical Approaches in Maritime Archaeology: A Mediterranean Perspective. *Antiquity* 64:376–389.

Gibbins, David, and Jonathan Adams

2001 Shipwrecks and Maritime Archaeology. *World Archaeology* 32(3):279–291.

Gibbs, Jim

1977 *Shipwrecks in Paradise: An Informal Marine History of the Hawaiian Islands*. Superior Publishing, Seattle.

Gibbs, Martin

1995 The Historical Archaeology of Shore-Based Whaling in Western Australia 1836–1879. Unpublished PhD dissertation, University of Western Australia, Perth.

2006 Cultural Site Formation Processes in Maritime Archaeology: Disaster Response, Salvage and Muckelroy Thirty Years On. *International Journal of Nautical Archaeology* 35(1):4–19.

2010 *The Shore Whalers of Western Australia: Historical Archaeology of a Maritime Frontier*. Vol. 2. Studies in Australasian Historical Archaeology. University of Sydney, Sydney, Australia.

Gibson, Arrell M., and John S. Whitehead

1993 *Yankees in Paradise: The Pacific Basin Frontier*. University of New Mexico Press, Albuquerque.

Gifford, K. Dun

1998 Meals Onboard Whaleships: Edible? Inedible? Incredible? In *Fish: Food from the Waters, Proceedings of the Oxford Symposium on Food and Cookery 1997*, edited by Harlan Walker, pp. 120–149. Devon, United Kingdom.

Gilbert, Glyndwr

1926 *The Death of Captain Cook*. Hawaiian Historical Society Reprints Number 5. Paradise of the Pacific Press, Honolulu.

Gill, James C. H.

1966 The Genesis of the Australian Whaling Industry. *Journal of the Royal Historical Society of Queensland* 8(1):111–136.

Gjerde, Jan M.

2010 Rock Art Landscapes: Studies in Stone Age Rock Art from Northern Fennoscandia. Unpublished PhD dissertation, University of Tromsø, Tromsø, Norway.

Gleason, Kelly

2008 *Activity Report: Maritime Heritage Resources Survey HA-08–04*. National Oceanic and Atmospheric Administration and Papahanaumokuakea Marine National Monument.

Godoy, Ricardo

1984 Mining: Anthropological Perspectives. *Annual Review of Anthropology* 14:199–217.

Goode, George B.

1887 *The Fisheries and Fishery Industries of the United States*. Vol. 2. US Government Printing Office, Washington, DC.

Gordon, Robert B., and Patrick M. Malone

1994 *The Texture of Industry: An Archaeological View of the Industrialization of North America*. Oxford University Press, New York.

Gosho, Merrill E., Dale W. Rice, and Jeffrey M. Breiwick

1984 The Sperm Whale, *Physeter macrocephalus*. *Marine Fisheries Review* 46(4): 54–64.

Gould, Richard A.

1997 *Archaeology and the Social History of Ships*. Cambridge University Press, Cambridge, United Kingdom.

1998 Shipwreck Anthropology. In *Encyclopedia of Underwater and Maritime Archaeology*, edited by James P. Delgado, pp. 377–380. Yale University Press, New Haven, Connecticut.

2011 *Archaeology and the Social History of Ships*. 2nd ed. Cambridge University Press, Cambridge, United Kingdom.

Gould, Richard A. (editor)

1983 *Shipwreck Anthropology*. University of New Mexico Press, Albuquerque.

Grant, Gordon

1932 *Greasy Luck: A Whaling Sketchbook*. Dover Publications, Mineola, New York.

Green, Cathy, and Jason T. Raupp.

2012 *HI12–03 Maritime Heritage Notes*. Papahanaumokuakea Marine National Monument.

Grenier, Robert

1988 Basque Whalers in the New World: The Red Bay Wrecks. In *Ships and Shipwrecks of the Americas*, edited by George F. Bass, pp. 69–84. Thames and Hudson, London.

Grenier, Robert, Marc-Andre Brenier, and Willis Stevens (editors)
2007 *The Underwater Archaeology of Red Bay: Basque Shipbuilding and Whaling in the 16th Century*. Vol. 5. Parks Canada, Ottawa.

Griffiths, John, and Kathy Mack
2011 Senses of "Shipscapes": An Artful Navigation of Ship Architecture and Aesthetics. *Journal of Organizational Change Management* 24(6):733–750.

Gschaedler, Andre
1948 Religious Aspects of the Spanish Voyages in the Pacific during the Sixteenth Century and the Early Part of the Seventeenth. *Americas* 4(3):302–315.

Gwyn, David
2007 *Gwynedd: Inheriting a Revolution; the Archaeology of Industrialisation in North-West Wales*. Philimore, Chichester, United Kingdom.

Hacquebord, Louwrens
1984 The History of Early Dutch Whaling: A Study from the Ecological Angle. In *Arctic Whaling: Proceedings of the International Symposium Arctic Whaling 1983*, edited by Hugo J. s'Jacob, Kim Snoeijing, and Richard Vaughn, pp. 135–148. Arctic Center University of Groningen, Groningen, Netherlands.
1987 A Historical-Archaeological Investigation of a Seventeenth-Century Whaling Settlement on the West Coast of Spitsbergen in 79° North Latitude. *Smeerenburg Seminar* 38:19-34.

Hall, Henry
1884 *Report on the Ship-building Industry of the United States*. US Government Printing Office, Washington, DC.

Hall, Martin, and Stephen W. Silliman (editors)
2006 Introduction: Archaeology of the Modern World. In *Historical Archaeology*, pp. 1–19. Blackwell, Malden, Massachusetts.

Hardesty, Donald L.
1988 *Mining Archaeology in the American West: A View from the Silver State*. Society for Historical Archaeology, Pleasant Hill, California.
2010 *The Archaeology of Mining and Mining Sites: A View from the Silver State*. University of Nebraska Press, Lincoln.

Harwood, P. L.
1935 *The Development of Whaling Implements*. Marine Historical Association, Mystic, Connecticut.

Hawaiian Gazette [Honolulu]
1868 Dissolution of the Firm of J. Robinson & Co. September 16, 4(35):2.

Hawaiian Spectator
1838 Ocean Island. 1(3):336–337.

Hawaii State Archives
1859 Pearl and Hermes Lat. '27° 42′ N. Long. 175°. 48.' File number G4382 L37 1859 B7.

Hawes, Charles B.
1924 *Whaling*. Doubleday, Page, Garden City, New York.

Hay, Alan

2009 Economic Time, Technological Space: A Spatial Analysis of Capitalism at the Phoenix Foundry 1847–1928. Unpublished PhD dissertation, Department of Anthropology, Flinders University.

Hazen, Jacob A.

1854 *Five Years before the Mast; or, Life in the Forecastle, aboard of a Whaler and Man-of-War.* Willis P. Hazard, Philadelphia.

Heffer, Jean

2002 *The United States and the Pacific: History of the Frontier.* University of Notre Dame Press, Notre Dame, Indiana.

Heffernan, Thomas F.

1981 *Stove by a Whale: Owen Chase and the* Essex. Wesleyan University Press, Middleton, Connecticut.

Hegarty, Reginald B.

1960 *New Bedford and American Whaling.* Reynolds Printing, New Bedford, Massachusetts.

1964 *The Birth of a Whaleship.* New Bedford Free Public Library, New Bedford, Massachusetts.

Hellman, Robert

2011 Some Remarks about Nantucket Whalecraft Makers. *Historic Nantucket* 61(3):15–20.

Henderson, Graeme

1983 *The Rowley Shoals Shipwreck Site: A Progress Report.* Western Australian Museum, Department of Maritime Archaeology Report, Fremantle, Australia.

1989 *Belinda.* Western Australian Maritime Museum, Department of Maritime Archaeology, Report.

2007 *Unfinished Voyages: Western Australian Shipwrecks 1622–1850.* 2nd ed. University of Western Australia Press, Crawley.

Hezel, Francis X.

1983 *The First Taint of Civilization: A History of the Caroline and Marshall Islands in Pre-Colonial Days, 1521–1885.* University of Hawaii Press, Honolulu.

Higgins, Joseph T.

1927 *The Whaleship Book.* Rudder Publishing, New York.

Hocker, Frederick M.

2004 Shipbuilding: Philosophy, Practice and Research. In *The Philosophy of Shipbuilding*, edited by Frederick M. Hocker and Cheryl Ward, pp. 1–11. Texas A&M Press, College Station.

Hodder, Ian

1987 *The Archaeology of Contextual Meanings.* Cambridge University Press, United Kingdom.

Hogue, C. E.

1947 The Whalers. *Hawaiian Digest*, pp. 7–8.

Hohman, Elmo P.

1926 Wages, Risk, and Profits in the American Whaling Industry. *Quarterly Journal of Economics* 40(4):644–671.

1928 *The American Whaleman*. Longmans, Green, New York.

Hosking, William J.

1973 Whaling in South Australia 1837–1872. Unpublished master's thesis, Department of Anthropology, Flinders University, Adelaide, Australia.

Hosty, Kieran, and Iain Stuart

1994 Maritime Archaeology over the Last Twenty Years. *Australian Archaeology* 39(1):9–19.

Howard, Mark

1996 Coopers and Casks in the Whaling Trade 1800–1850. *Mariner's Mirror* 82(4):436–450.

Howay, F. W.

1932 An Outline Sketch of the Maritime Fur Trade: Presidential Address. *Report of the Annual Meeting of the Canadian Historical Association* 11(1):5–14.

Howe, Kerry R.

1984 *Where the Waves Fall: A New South Sea Islands History from Frontier Settlement to Colonial Rule*. University of Hawaii Press, Honolulu.

Hume, Ivor N.

1991 *A Guide to Artifacts of Colonial America*. Vintage Books, New York.

Hussey, Edwin

n.d. Plan of Stowage in Lower Hold in Ship *Gratitude* by Edwin Hussey 1 mo 28. 41. MSS 78, Sub-group 3, Series J, Sub-series 1, Folder 2. New Bedford Whaling Museum Research Library and Archives, New Bedford, Massachusetts.

Hyde, Alexander, Abraham C. Baldwin, and William L. Gage

1874 *The Frozen Zone and Its Explorers: A Comprehensive Record of Voyages, Travels, Discoveries, Adventures, and Whale Fishing in the Arctic Regions for One Thousand Years*. Colombian Book, Hartford, Connecticut.

Hydrographic Office

1903 *The Hawaiian Islands and the Islands, Rocks, and Shoals to the Westward*. 2nd ed. US Government Printing Office, Washington, DC.

Igler, David

2004 Diseased Goods: Global Exchanges in the Eastern Pacific Basin, 1770–1850. *American Historical Review* 109(3):693–719.

Ironbridge Gorge Museum Trust

2013 Objects That Changed the World. Accessed November 16, 2013.

Isachsen, Gunnar

1929 Modern Norwegian Whaling in the Antarctic. *Geographical Review* 19(3):387–403.

Israel, Jonathan S.

1990 *Dutch Primacy* in *World Trade, 1585–1740*. Oxford University Press, New York.

Jackson, Gordon
1978 *The British Whaling Trade*. Archon Books, Hamden, Connecticut.

Jefferson, Thomas
1788 Observations on the Whale Fishery. In *Thomas Jefferson, Writings*. Jacques Gabriel Clousier, Paris.

Jenkins, J. Travis
1921 *A History of the Whale Fisheries: From the Basque Fisheries of the Tenth Century to the Hunting of the Finner Whale at the Present Date*. H. F. & G. Witherby, London.

Jobling, Harold J. W.
1993 The History and Development of English Anchors 1550–1850. Unpublished master's thesis, Department of Anthropology, Texas A&M University, College Station.

Johnson, Donald D.
1995 *The United States in the Pacific: Private Interests and Public Policies 1784 1899*. Praeger Publishers, Westport, Connecticut.

Johnson, Matthew
1996 *The Archaeology of Capitalism*. Blackwell Publishing, Oxford, United Kingdom.

Jones, Alfred. G. E.
1981a The British Southern Whale and Seal Fisheries. *The Great Circle* 3(1):20–29.
1981b The British Southern Whale and Seal Fisheries—Part II: The Principal Operators. *Great Circle* 3(2):90–102.
1986 *Ships Employed in the South Seas Trade 1775–1859*. Vol. 2. Roebuck Society Publication, Victoria, Canada.

Jones, Toby N.
2004 The Mica Shipwreck: Deepwater Nautical Archaeology in the Gulf of Mexico. Unpublished master's thesis, Department of Anthropology, Texas A&M University, College Station.

Kaplan, Sidney
1953 Lewis Temple and the Hunting of the Whale. *New England Quarterly* 26(1):78–88.

Kenderdine, S.
1994 Lancier *(1834–1839) Wreck Inspection Report*. Western Australian Maritime Museum, Fremantle, Australia.

Kenney, Robert D., Charles A. Mayo, and Howard E. Winn
2001 Migration and Foraging Strategies at Varying Spatial Scales in Western North Atlantic Right Whales: A Review of Hypotheses. *Journal of Cetacean Resource Management* 2:251–60.

Kimpton, Geoff, and Graeme Henderson
1991 The Last Voyage of the *Day Dawn* Wreck. *Bulletin of the Australian Institute for Maritime Archaeology* 15(2):25–28.

King, Robert D. (editor)
1931 *Index to the Islands of the Territory of Hawaii: Including Other Islands under*

the Sovereignty of the United States Scattered in the North Pacific Ocean. US Government Printing Office, Washington, DC.

Kipping, Robert

1859 *Rudimentary Treatise on Masting, Mast-Making, and Rigging of Ships*. John Weale, London.

Kirby, H. S.

1860 *Memorandum of Whaler's Outfits*. Published by author, New Bedford, Massachusetts.

Kirch, Patrick

2011 When Did the Polynesians Settle Hawai'i? A Review of 150 Years of Scholarly Inquiry and a Tentative Answer. *Hawaiian Archaeology* 12:3–26.

Kittinger, John N.

2010 Historical Ecology of Coral Reefs in the Hawaiian Archipelago. Unpublished PhD dissertation, Department of Anthropology, University of Hawaii at Manoa.

Kittinger, John N., A. Dowling, A. R. Purves, N. A. Milne, and P. Olsson

2011 Marine Protected Areas, Multiple-Agency Management, and Monumental Surprise in the Northwestern Hawaiian Islands. *Journal of Marine Biology* 2011:1–17.

Kittinger, John N., K. N. Duin, and B. A. Wilcox

2010 Commercial Fishing, Conservation and Compatibility in the Northwestern Hawaiian Islands. *Marine Policy* 34:208–217.

Knapp, Arthur

1998 Social Approaches to the Archaeology and Anthropology of Mining. In Knapp et al. *Social Approaches to an Industrial Past*, pp. 1–23.

Knapp, Arthur, Vincent C. Piggot, and Eugenia W. Herbert (editors)

1998 *Social Approaches to an Industrial Past: The Archaeology and Anthropology of Mining*. Routledge, London.

Kugler, Richard C.

1971 The Penetration of the Pacific by American Whalemen in the 19th Century. In *The Opening of the Pacific—Image and Reality*. Maritime Monographs and Reports 2:20–27. National Maritime Museum, Greenwich, United Kingdom.

1980 The Whale Oil Trade 1750–1775. *Old Dartmouth Historical Sketch* 79.

1981 Foreword. In Davoll, *Captain's Specific Orders*.

Kuykendall, Ralph S.

1931 American Interests and American Influence in Hawaii in 1842. *Hawaiian Historical Society Thirty-Ninth Annual Report for the Year*, pp. 48–67.

1934 Early Hawaiian Commercial Development. *Pacific Historical Review* 3(4):365–385.

Laing, John A.

1815 *A Voyage to Spitzbergen: Containing an Account of that Country, of the Zoology of the North, of the Shetland Isles, and of the Whale Fishery*. J. Mawman, London.

Lawrence, Susan, and Mark Staniforth

1998 The Archaeology of Whaling in Southern Australia and New Zealand. *Australasian Society for Historical Archaeology and the Australian Institute for Maritime Archaeology Special Publication* 10.

Leavitt, John F.

1973 *The Charles W. Morgan*. Marine Historical Association, Mystic Seaport, Connecticut.

Lebo, Susan A.

2007 Native Hawaiian Whalers in Nantucket, 1820–60. *Historic Nantucket* 56(1):14–16.

Lee, Ki-baik

1984 *A New History of Korea*. Harvard University Press, Cambridge, Massachusetts.

Leff, David N.

1940 *Uncle Sam's Pacific Islets*. Stanford University Press, California.

Leslie, John, Robert Jameson, and Hugh Murray

1850 *Narrative of Discovery and Adventure in the Polar Seas and Regions: With Illustrations of Their Climate, Geology and Natural History; and an Account of the Whale Fishery*. Oliver & Boyd, Edinburgh, United Kingdom.

Light, John D.

1990 The 16th Century Anchor from Red Bay, Labrador: Its Method of Manufacture. *International Journal of Nautical Archaeology and Underwater Exploration* 19(4):307–316.

1992 16th Century Basque Ironworking: Anchors and Nails. *Materials Characterization* 29:249–258.

2007 A Dictionary of Blacksmithing Terms. *Historical Archaeology* 41(2):84–157.

Lindquist, Oie

1993 Whaling by Peasant Fishermen in Norway, Orkney, Shetland, the Faroe Islands, Iceland and Norse Greenland: Mediaeval and Early Modern Whaling Methods and Inshore Legal Régimes. In *Whaling and History: Perspectives on the Evolution of the Industry*, edited by Jan E. Ringstad, Einar Wexelsen, and Bjørn L. Basberg, pp. 81–90. Kommadør Chr. Christiansens Hvalfangstmuseum, Sandefjorde, Norway.

Lindsey, Bill

2023 BLM/SHA Historic Bottle Identification and Information (website). Accessed August 23, 2023.

Lipman, Arthur G., and Geo E. Osborne

1969 Medicine and Pharmacy aboard New England Whaleships. *Pharmacy in History* 11(4):119–131.

Little, Barbara

1969 The Sealing and Whaling Industry in Australia before 1850. *Australian Economic History Review* 9(2):109–127.

1994 People with History: An Update on Historical Archaeology in the United States. *Journal of Archaeological Method and Theory* 1(1):5–40.

Littlefield, L. A.
1906 Fitting Out a Whaler. *Old Dartmouth Historical Sketch* 14:4–13.

Logan, Judith A., and James A. Tuck
1990 A Sixteenth Century Basque Whaling Port in Southern Labrador. *APT Bulletin* 22(3):65–72.

Lubanova, N. V.
2007 Petroglyphs of Staraya Zalavruga: New Evidence—New Outlook. *Archaeology, Ethnography and Anthropology of Eurasia* 29(1):127–135.

Luce, Stephen B.
1863 *Seamanship: Compiled from Various Authorities and Illustrated with Numerous and Selected Designs for the Use of the United States Naval Academy*. 2nd ed. James Atkinson Printer, Newport, Rhode Island.

Lydgate, J. M.
1915 Wrecks to the Northwest. In *The Hawaiian Annual and Almanac: The Reference Book of Hawaii*, edited by Thomas G. Thrum, pp. 133–144. Thrum Publisher, Honolulu, Hawaii.

Lytle, Thomas G.
1984 *Harpoons and Other Whalecraft*. New Bedford Whaling Museum, New Bedford, Massachusetts.

MacAllan, Richard
2000 Entrepôt to the World: Richard Charlton's Observations of Trade via Hawai'i, 1828–1841. *Hawaiian Journal of History* 34:93–111.

MacGregor, David R.
1989 *Merchant Sailing Ships, 1815–1850: Supremacy of Sail*. Vol. 2. 2nd ed. Naval Institute Press, Annapolis, Maryland.

Macknight, Charles C.
1976 *The Voyage to Marege': Macassan Trepangers in Northern Australia*. Melbourne University Press, Australia.

Macy, Obed
1835 *History of Nantucket and of the Whale Fishery*. Hilliard, Gray, Boston.

Macy, William H.
1877 *Thar She Blows! Or, the Log of the Arethusa*. Lee and Shepard, Boston.

Manchester, Curtis A., Jr.
1951 The Exploration and Mapping of the Pacific. In *Geography of the Pacific*, edited by O. W. Freeman, pp. 61–88. John Wiley & Sons, New York.

Maragos, James E., and David Gulko
2002 *Coral Reef Ecosystems of the Northwestern Hawaiian Islands: Interim Results Emphasizing the 2000 Surveys*. US Fish and Wildlife Service and the Hawaii Department of Land and Natural Resources, Honolulu.

Martin, Jay C.
2014 Strands That Stand: Using Wire Rope to Date and Identify Archaeological Sites. *International Journal of Nautical Archaeology* 43(1):151–161.

Martin, Kenneth R.
1974 *Delaware Goes Whaling, 1833–1845*. Hagley Museum, Greenville, Delaware.

1975 *Whalemen and Whaleships of Maine: A Publication of the Marine Research Society of Bath, Maine.* Harpswell Press, Brunswick, Maine.

Martin, Patrick E.

2009 Industrial Archaeology. In *International Handbook of Historical Archaeology*, edited by David Gaimster and Teresita Majewski, pp. 285–297. Springer, New York.

Mate, Geraldine

2010 Mining the Landscape: Finding the Social in the Industrial through an Archaeology of the Landscapes of Mount Shamrock. Unpublished PhD dissertation, Department of Anthropology, University of Queensland, Brisbane, Australia.

2013 Mount Shamrock: A Symbiosis of Mine and Settlement. *International Journal of Historical Archaeology* 17(3):465–486.

Maui Historical Society

1964 *Lahaina Historical Guide.* 2nd ed. Charles E. Tuttle, Tokyo.

Maury, Matthew F.

1851 Whale Chart, Color Lithograph. G9096s. C7 var. M3, series F (8). Naval Observatory, Geography and Map Division.

Maury, Nannie B.

1896 *Whalers and Whaling.* H. S. Hutchinson, New Bedford, Massachusetts.

Mawer, Granville A.

1999 *Ahab's Trade: The Saga of South Seas Whaling.* Allen & Unwin, St. Leonards, Australia.

McAllister, Madeline

2012 *Koombana Bay Foreshore Excavation–Structural Analysis.* Department of Maritime Archaeology, Western Australian Museum, Perth.

2013 Stout, Sturdy and Strong: A Typology for Early Nineteenth-Century American Whalers. Unpublished PhD dissertation, Department of Anthropology, Flinders University, Adelaide, Australia.

McCarthy, Michael

1981 American Whaler Excavated in Cockburn Sound (Western Australia). *Great Circle* 3(2):132–137.

1982 *Koombana Bay Wrecks: An Investigation of Wrecks in the Bay for the State Energy Commission of Western Australia.* Report on file, Department of Maritime Archaeology, Western Australian Museum, Perth.

1998 Australian Maritime Archaeology: Changes, Their Antecedents and the Path Ahead. *Australian Archaeology* 47(1):33–38.

2005 *Ships' Fastenings: From Sewn Boat to Steamship.* Texas A&M University Press, College Station.

McCartney, Allen P.

1984 History of Native Whaling in the Arctic and Subarctic. In *Arctic Whaling: Proceedings of the International Symposium Arctic Whaling*, edited by Kim Snoeijing and Hugo J. s'Jacob, pp. 79–111. Arctic Center University of Groningen, Groningen, Netherlands.

McConnell, Kenneth E., and Michael Price

2006 The Lay System in Commercial Fisheries: Origin and Implications. *Journal of Environmental Economics and Management* 51(3):295–307.

McDowell Ward, Heather E.

2010 Creating the Papahanaumokuakea Marine National Monument: Discourse, Media, Place-Making, and Policy Entrepreneurs. Unpublished master's thesis, Department of History, East Carolina University, Greenville, North Carolina.

McGann, Sally

1999 Wilyah Mia: An Archaeology of the Shark Bay Pearling Industry 1850–1930. Unpublished master's Thesis, Centre for Archaeology, University of Western Australia, Perth.

McGowan, Barry

2003 The Archaeology of Chinese Alluvial Mining in Australia. *Australasian Historical Archaeology*, 11–17.

McKinnon, Jennifer, Julie Mushynsky, and Genevieve Cabrera

2014 A Fluid Sea in the Mariana Islands: Community Archaeology and Mapping the Seascape of Saipan. *Journal of Maritime Archaeology* 9(1):59–79.

McVarish, Douglas C.

2008 *American Industrial Archaeology: A Field Guide*. Left Coast Press, Walnut Creek, California.

Meares, John

1791 *Voyages Made in the Years 1788 and 1789, from China to the NW Coast of America: With an Introductory Narrative of a Voyage Performed in 1786, from Bengal, in the Ship* Nootka. *To Which Are Annexed, Observations on the Probable Existence of a North West Passage. And Some Account of the Trade between the North West Coast of America and China; and the Latter Country and Great Britain*. Logographic Press, London.

Meide, Chuck, S. P. Turner, B. Burke, and Starr Cox

2011 *First Coast Maritime Archaeology Project 2010: Report on Archaeological Investigations*. Copies available from State of Florida Bureau of Archaeological Research, St. Augustine.

Mellor, Ian

2005 Space, Society and the Textile Mill. *Industrial Archaeology Review* 27(1):49–56.

Melville, Herman

1851 *Moby Dick; or, The Whale*. Harper & Brothers, New York.

Miller, David G.

1988 Ka'iana, the Once Famous "Prince of Kaua'i." *Hawaiian Journal of History* 22:1–19.

Moran, Geoffrey P.

1976 Trash Pits and Natural Rights in the Revolutionary Era: Excavations at the Narbonne House in Salem, Massachusetts. *Archaeology* 29(3):194–202.

Moran, Vivienne

1997 Some Management Options for the Perched Hull *Day Dawn*. *Bulletin of the Australian Institute for Maritime Archaeology* 21:129.

Morgan, T.

1948 *Hawaii: A Century of Economic Change, 1778–1876*. Harvard University Press, Cambridge, Massachusetts.

Morrell, B., Jr

1832 *A Narrative of Four Voyages to the South Sea, North and South Pacific Ocean, Chinese Sea, Ethiopic and Southern Atlantic Ocean, Indian and Antarctic Ocean*. Harper and Brothers, New York.

Morrison, Samuel E.

1921 *The Maritime History of Massachusetts 1783–1860*. Houghton Mifflin, Boston.

Morton, Harry

1982 *The Whale's Wake*. University of Hawaii Press, Honolulu.

Mrantz, Maxine

1976 *Whaling Days in Old Hawaii*. Tong Publishing, Honolulu, Hawaii.

Muckelroy, Keith

1978 *Maritime Archaeology*. Cambridge University Press, Cambridge, United Kingdom.

Munsey, Cecil

1970 *The Illustrated Guide to Collecting Bottles*. Hawthorne Books, New York.

Nantucket Historical Association Research Library (NHA)

1807 Macy "New Ship" Outfitting Book.

1820 Ship *Peru* Account Book.

1834 Shipbuilding Contract.

Nash, Michael

1990 Survey of the Historic Ship *Litherland* (1834–1853). *Bulletin of the Australian Institute for Maritime Archaeology* 14(1):13.

2001 *Cargo for the Colony: The 1797 Wreck of the Merchant Ship* Sydney Cove. Navarine Publishing, Woden, Australia.

2003 *The Bay Whalers: Tasmania's Shore-Based Whaling Industry*. Navarine Publishing, Woden, Australia.

2009 Sydney Cove*: The History and Archaeology of an Eighteenth-Century Shipwreck*. Navarine Publishing, Hobart, Australia.

Naval Intelligence Division

1945 *Pacific Islands: 1. General Survey*. Geographical Handbook Series. US Navy, Washington, DC.

Neumann, George

1984 *Early American Antique Country Furnishings: Northeastern America, 1650–1800*. American Legacy Press, New York.

Nevell, Michael

2006 The 2005 Rolt Memorial Lecture Industrial Archaeology or the Archaeology of the Industrial Period? Models, Methodology and the Future of Industrial Archaeology. *Industrial Archaeology Review* 28(1):3–15.

New Bedford Free Public Library Whaling Collection Archives (NBFPL)

1840 Bark *Marcella*, Outfitting Book.

1841 Bark *Canton*, Outfitting Book.

1845 Bark *Mars*, Outfitting Book.

1851 Ship *William C. Nye*, Outfitting Book.

New Bedford Whaling Museum Research Library and Archives (NBWM)

1800 Craft, Tools etc for a 3 Boat Ship—ca. 1800. MSS 56, box 39, ser. H, sub-ser. 17.

1819 *William Rotch* Outfitting Records.

1820 Of Provisions and Whalecraft Needed to Outfit a Whaler of 100–200 Tons for a Voyage of 12 Months—ca. 1820. MSS 56, box 39, ser. H, sub-ser. 23.

1832 *Condor* (Ship) Records, 1831–1850, Outfitting Book ca. 1832. MSS 41, box 1, sub-group 1, ser. E, vol. 1.

n.d. Cask stowage plan for unidentified ship. MSS 78, sub-group 3, ser. G, sub-ser. 1, folder 2.

Newman, Richard

2001 Landscape Archaeology. In *The Historical Archaeology of Britain, c. 1540–1900*, edited by Richard Newman, David Cranstone, and Christine Howard-Davis, pp. 100–182. Routledge, London.

New York Times

1975 Whaleship *Ansel Gibbs* to Be Recovered. April 27:217.

Nickerson, T.

n.d. Loss of the Ship *Two Brothers* of Nantucket. MS 106, folder 3.5. Nantucket Historical Association Research Library.

1876 Letter to Leon Lewis, Esq. NHA MS 106. Nantucket Historical Association Research Library.

Noel, S. B. J.

1809 An Historical Memoire, Relative to the Antiquity of the Whale Fishery. *Monthly Magazine or British Register* 27(1):690–693.

Nordhoff, Charles

1874 *Whaling and Fishing*. Moore, Wilstach, Keys, Cincinnati, Ohio.

Norling, Lisa

1996 Ahab's Wife: Women and the American Whaling Industry, 1820–1870. In *Iron Men, Wooden Women: Gender and Seafaring in the Atlantic World, 1700–1920*, edited by Lisa Norling and Margaret S. Creighton, pp. 70–91. Johns Hopkins University Press, Baltimore, Maryland.

North American Review

1834 The Whale Fishery. 38(82):84–116.

Norton, L. A.

2012 The Pacific in the War of 1812: Pelts, Ploys and Plunder. *Coriolis* 3(1):1–20.

Nutley, David

1987 Investigations into an Early Eighteenth Century Instrument Believed to Be a Set of Dividers from the Rowley Shoals Wreck. *Bulletin of the Australian Institute for Maritime Archaeology* 11(2):41.

Oliver, Douglas L.
1989 *The Pacific Islands*. 3rd ed. University of Hawaii Press, Honolulu.

Olly, J.
2004 *Lagoda*: A Legacy in Wood, Iron, and Sail. *Nautical Research Journal* 49:144–153.

Olmsted, Francis A.
1841 *Incidents of a Whaling Voyage: To Which Are Added Observations on the Scenery Manners and Customs, and Missionary Stations of the Sandwich and Society Islands: Accompanied by Numerous Lithographic Plates*. Reprint 1969 by Bell Publishing, New York.

Ommanney, Francis D.
1971 *Lost Leviathan*. Hutchinson, London.

Orange, Hilary
2008 Industrial Archaeology: Its Place within the Academic Discipline, the Public Realm and the Heritage Industry. *Industrial Archaeology Review* 30(2):83–95.

Orser, Charles E., Jr.
2004 *Historical Archaeology*. 2nd ed. Pearson Education, Upper Saddle River, New Jersey.

Paasch, Heinrich
1885 *Paasch's Illustrated Marine Dictionary: In English, French, and German, Originally Published as from Keel to Truck, Reprint 1997.* Conway Maritime Press, London.

Pacific Commercial Advertiser [Honolulu, Hawaii]
1859 "Loss of the American Whale Ship *South Seaman*." March 31, 3(40):2.
1867a "Loss of the Bark *Daniel Wood*." April 27, 11(43):2.
1867b "Wrecking." May 18, 11(46):3.

Paddack, William C.
1893 *Life on the Ocean: Or Thirty Five Years at Sea. Being the Personal Adventures of the Author*. Riverside Press, Cambridge, United Kingdom.

Palmer, M., and P. Neaverson
1998 *Industrial Archaeology: Principles and Practice*. Routledge, London.

Palmer, Marilyn
1990 Industrial Archaeology: A Thematic or a Period Discipline? *Antiquity* 64(243):275–282.
2005 Understanding the Workplace: A Research Framework for Industrial Archaeology in Britain. *Industrial Archaeology Review* 27(1):9–17.

Palmer, W. R.
1959 The Whaling Port of Sag Harbor. Unpublished PhD dissertation, Department of History, Columbia University, New York.

Pandolfi, John M., Jeremy B. C. Jackson, N. Baron, Roger H. Bradbury, Hector M. Guzman, Terence P. Hughes, C. V. Kappel et al.
2005 Are U.S. Coral Reefs on the Slippery Slope to Slime? *Science* 307:1725–1726.

Papahanaumokuakea Marine National Monument (PMNM)

2008 *Papahānaumokuākea Marine National Monument Management Plan.* Honolulu, Hawaii.

2011 *Maritime Heritage Research, Education, and Management Plan: Papahānaumokuākea Marine National Monument.* Honolulu, Hawaii.

Patterson, Adam

2006a If It Ain't Broke—Why Fix it? A Comparative Historical and Archaeological Study of the Technological Development of Tryworks at Shore Based Whaling Stations, 1550–1871. In *National Archaeology Students Conference: Explorations, Investigations and New Directions*, edited by D. Arthur and Adam Paterson, pp. 59–69. Flinders Press, Bedford Park, Australia.

2006b *The Sleaford Bay Tryworks: Industrial Archaeology of Shore Based Whaling Stations.* Flinders University Maritime Archaeology Monographs Series 3. Shannon Research Press, Adelaide, Australia.

Payne, C. W.

n.d. Notes on Whaling. John Jermain Memorial Library, Sag Harbor, New York.

Pearson, Michael

1983 The Technology of Whaling in Australian Waters in the 19th Century. *Australian Journal of Historical Archaeology*, 40–54.

Pease, Zephaniah W.

1916 The Story of the Building of the Bourne Whaling Museum. *Old Dartmouth Historical Sketches* 44:13–22.

Perkins, Edward T.

1854 *Na Motu or, Reef-Rovings in the South Seas: A Narrative of Adventures at the Hawaiian, Georgian, and Society Islands.* Pudney and Russell, New York.

Pfaffenberger, Bryan

1998 Mining Communities, Chaînes Opératoires and Sociotechnical Systems. In Knapp et al., *Social Approaches to an Industrial Past*, pp. 291–300.

Philbrick, Nathaniel

2001 *In the Heart of the Sea: The Tragedy of the Whaleship* Essex. Penguin Books, London.

Pollard, G., Jr.

1834 Narrative of Loss of *Essex* and Rescue by Ship *Dauphin*. MS 106. "Ship *Peru*'s Book 1820." AB 267. Hawaii State Archives.

Polynesian [Honolulu, Hawaii]

1844 Shipwreck of the *Holder Borden*, American Whaler. October 12, 1(21):3.

1852 Narrative of the Loss of the Ship *Huntress*—Shipwreck and Suffering of the Crew. October 23, 9(24):1.

Poole, Dorothy C.

1977 The Captain's Medicine Chest. Dukes County Intelligencer 19(1):3–18.

Porthouse, Thomas

1848 *The Chronometer: Its Origin, and Present Perfection.* Poplett, London.

Potter, Parker B., Jr.

1990 An Envelope Full of Questions That Count in Underwater Archaeology. In *Underwater Archaeology Proceedings from the Society for Historical Archaeology*

Conference 1990, edited by Toni L. Carrell, pp. 34–38. Society for Historical Archaeology, Tucson, Arizona.

Pratt, A.

1859 Testimony of Captain Thos. Norton, Wreck of *South Seaman*. In *Consulate of the United States of America at Honolulu*. Consulate of the United States of America at Honolulu, Hawaii.

Proulx, Jean-Pierre

1986 *Whaling in the North Atlantic: From Earliest Times to the Mid-19th Century*. Parks Canada, Ottawa.

2007 Basque Whaling Methods, Technology and Organization in the 16th Century. In *The Underwater Archaeology of Red Bay. Basque Shipbuilding and Whaling in the 16th Century*, edited by R. Grenier, M. A. Brenier, and W. Stevens, pp. 42–96. Parks Canada, Ottawa.

Pūkui, Mary K., Samuel H. Elbert, and Esther T. Mookini

1974 *Place Names of Hawaii*. University Press of Hawaii, Honolulu.

Quivik, Fredric L.

2000 Landscapes as Industrial Artifacts: Lessons from Environmental History. *Journal of the Society for Industrial Archeology*, 55–64.

Ralston, Caroline

1978 *Grass Huts and Warehouses: Pacific Beach Communities of the Nineteenth Century*. University of Hawaii Press, Honolulu.

Raper, Henry

1840 *The Practice of Navigation and Nautical Astronomy*. R. B. Bate, London.

Raupp, Jason T.

2004 Hook, Line, and Sinker: Historical and Archaeological Investigations of the Snapper Wreck (8SR1001). Unpublished master's thesis, Department of Anthropology, University of West Florida, Pensacola.

Raupp, Jason T., and Kelly Gleason

2010 Submerged Whaling Heritage in Papahanaumokuakea Marine National Monument. *Journal of the Australasian Institute for Maritime Archaeology* 34:66–74.

Rauzon, Mark J.

2001 *Isles of Refuge: Wildlife and History of the Northwestern Hawaiian Islands Honolulu*. University of Hawaii Press, Honolulu.

Read, George H.

1912 *The Last Cruise of the Saginaw*. Houghton Mifflin, Boston.

Redknap, Mark (editor)

1997 *Artefacts from Wrecks: Dated Assemblages from the Late Middle Ages to the Industrial Revolution*. Oxbow Books, Oxford, United Kingdom.

Reeves, Randall R., and Edward Mitchell

1986 The Long Island, New York, Right Whale Fishery: 1650–1924. *Report of the International Whaling Commission Special Issue* 10:201–220.

Reynolds, J. N.

1835 Information Collected by the Navy Department Relating to Islands, Reefs, Shoals, Etc., in the Pacific Ocean and South Seas, and Showing the

Expediency of an Exploring Expedition in that Ocean and Those Seas by the Navy. *American State Papers*, class 6, Naval Affairs, vol. 4:712. US Government Printing Office, Washington, DC.

Richards, Rhys

1994 *Into the South Seas: The Southern Whale Fishery Comes of Age on the Brazil Banks, 1765 to 1812: A Review of the Whaling Activities of American, British, French, Spanish and Portuguese Whalemen off Brazil and Patagonia before 1812.* Paramatta Press, Wellington, New Zealand.

1999 Honolulu and Whaling on the Japan Grounds. *American Neptune* 59(3):189–197.

Riley, Ray

2005 The Notions of Production and Consumption in Industrial Archaeology: Towards a Research Agenda. *Industrial Archaeology Review* 27(1):41–47.

Roberts, David H. (editor)

1992 *18th Century Shipbuilding. Remarks on the Navies of the English and Dutch from Observations Made at Their Dockyards in 1737, by Blaise Ollivier, Master Shipwright to the King of France.* Jean Boudroit Publications, East Sussex, United Kingdom.

Robotti, Frances D.

1955 Whalers as Explorers and Discoverers in the Pacific. *Nautical Research Journal* 7(1–2):23–25.

1962 *Whaling and Old Salem: A Chronicle of the Sea*. Fountainhead Publishers, New York.

Röding, Johann H.

1793 *Allgemeines Wörterbuch der Marine, in Allen Europaeischen Seesprachen*, Vol. 1. Gebauer, Hamburg, Germany.

Ronnberg, Erik A. R.

1985 *To Build a Whaleboat: Historical Notes and a Modelmaker's Guide*. New Bedford Whaling Museum, New Bedford, Massachusetts.

Ross, Lester A.

1985 16th-Century Spanish Basque Coopering. *Historical Archaeology* 19(1):1–31.

Rubin, Norman

1971 The Anchor. *Nautical Research Journal* 18(4):230–250.

Russell, Scott

2009 Prehistoric Settlement and Indigenous Watercraft of the Northern Mariana Islands. In *Maritime History and Archaeology of the Commonwealth of the Northern Mariana Islands, Report for Commonwealth of the Northern Mariana Islands Historic Preservation Office, Saipan*, edited by Toni Carrell, pp. 61–94. Saipan, Northern Mariana Islands.

Rydell, Raymond A.

1952 *Cape Horn to the Pacific*. University of California Press, Los Angeles.

Sanderson, Ivan T.

1956 *Follow the Whale*. Bramhall House, New York.

Sandwich Island Gazette [Honolulu, Hawaii]

1837 Another Disaster. November 11, 2:2.

Sanger, Chesley W.

1995 The Origins of British Whaling: Pre-1750 English and Scottish Involvement in the Northern Whale Fishery. *Northern Mariner* 5(3):15–32.

Saturday Press [Honolulu, Hawaii]

1883 Reminisces of Honolulu. April 28, 3(35):1.

Savelle, James M., and Nobuhiro Kishigami

2013 Anthropological Research on Whaling: Prehistoric, Historic and Current Contexts. *Senri Ethnological Studies* 84:1–48.

Scammon, Charles M.

1874 *The Marine Mammals of the North-western Coast of North America: Described and Illustrated; Together with an Account of the American Whale-fishery*. G. P. Putnam & Sons, New York.

Schmitt, Robert C., and Eleanor C. Nordyke

2001 Death in Hawai'i: The Epidemics of 1848–1849. *Hawaiian Journal of History* 35:1–13.

Schokkenbroek, Joost C. A.

2008 *Trying-Out: An Anatomy of Dutch Whaling and Sealing in the Nineteenth Century, 1815–1885*. Amsterdam University Press, Amsterdam, Netherlands.

Schug, Donald M.

2001 Hawai'i's Commercial Fishing Industry: 1820–1945. *Hawaiian Journal of History* 35:15–34.

Schultz, Charles R.

1967 Costs of Constructing and Outfitting the Ship *Charles W. Morgan*, 1840–1841. *Business History Review* 41(2):198–216.

Schwemmer, R.

2004 *"Shipwreck Database Summary:* Daniel Wood," *U.S. Pacific Coast Shipwreck Database.* Report on file, Papahanaumokuakea Marine National Monument, Honolulu, Hawaii.

Scoresby, William Jr.

1820 *An Account of the Arctic Regions, with a History and Description of the Northern Whale Fishery*. Archibald Constable, Edinburgh, United Kingdom.

SEARCH

2010 *2010 Marine Remote Sensing Survey*. Report on file, Papahanaumokuakea Marine National Monument, Monument, Honolulu, Hawaii.

Seward, William Henry

1852 *Commerce in the Pacific Ocean*. Buell & Blanchard, Washington, DC.

Shapiro, Irwin

1959 *The Story of Yankee Whaling*. American Heritage Publishing, New York.

Sharp, Andrew

1960 *The Discovery of the Pacific Islands*. Clarendon Press, Oxford, United Kingdom.

Shefi, Debra G.

2006 The Development of Cutters in Relation to the South Australian Oyster Industry: An Amalgamation of Two Parallel Developing Industries. Unpublished master's thesis, Department of Archaeology, Flinders University, Adelaide, Australia.

Sherman, Stuart C.

1965 *The Voice of the Whaleman: With an Account of the Nicholson Whaling Collection*. Providence Public Library, Providence, Rhode Island.

Shoemaker, James H.

1948 *Economy of Hawaii in 1947: With Special Reference to Wages, Working Conditions and Industrial Relations: Bulletin of the United States Bureau of Labor Statistics, No. 926*. US Government Printing Office, Washington, DC.

Shoemaker, Nancy

2015 *Native American Whalemen and the World: Indigenous Encounters and the Contingency of Race*. University of North Carolina Press, Chapel Hill.

Sievers, Burkard

2009 Before the Surrogate of Motivation: Motivation and the Meaning of Work in the Golden Age of the American Whaling Industry Part 1. *Critique: Journal of Socialist Theory* 37(3):391–413.

Silliman, Stephen

2006 Struggling with Labor, Working with Identities. In *Historical Archaeology*, edited by Martin Hall and Stephen Silliman, pp. 147–166. Blackwell Publishing, Malden, Massachusetts.

Silva, C. F., J. E. Bradford, and J. F. Canepa

1990 *KO-A4–3D-148 Lelu Harbour, Kosrae, FSM.* Report on file, Federated States of Micronesia Historic Preservation Office and Kosrae State Historic Preservation Office, Tofol, Kosrae, Federated States of Micronesia.

Skelton, Reginald

1954 Captain James Cook as a Hydrographer. *Mariner's Mirror* 40:91–119.

Skowronek, Russel K.

1998 The Spanish Philippines: Archaeological Perspectives on Economics and Society. *International Journal of Historical Archaeology* 2(1):45–71.

Smith, Derek

2010 The Ecology of Shipwrecks: An Assessment of Biodiversity. Unpublished master's thesis, Department of Zoology, University of Hawai'i at Manoa.

Smith, Ian

2002 *The New Zealand Sealing Industry: History, Archaeology, and Heritage Management*. Department of Conservation, Wellington, New Zealand.

Smith, Roger C.

2000 *The Maritime Heritage of the Cayman Islands*. University Press of Florida, Gainesville.

Smith, Roger C., Jeffrey T. Moates, Debra G. Shefi, Brian J. Adams, Brenda S. Altmeier, Lee A. Newsom, and Colleen L. Reese

2006 *Archaeological and Biological Examination of the Brick Wreck (8MO1881) off*

Vaca Key, Monroe County, Florida. Report on file, Florida Keys National Marine Sanctuary, Key Largo, Florida

Smithsonian Institution Museum of American History

2014 "On the Water" Electronic Exhibition. Smithsonian Institution Museum of American History (website), accessed October 9, 2014.

Smithsonian Institution Museum of Natural History

2014 Arctic Studies Center. "St. Lawrence Gateways Project." Smithsonian Institution Museum of Natural History (website), accessed October 3, 2014.

Songini, Marc

2007 *The Lost Fleet: A Yankee Whaler's Struggle against the Confederate Navy and Arctic Disaster*. St. Martin's Griffin, New York.

Souza, Donna J.

1998 *The Persistence of Sail in the Age of Steam: Underwater Archaeological Evidence from the Dry Tortugas*. Plenum Press, New York.

Spears, John R.

1910 *The Story of the New England Whalers*. 2nd ed. Macmillan, New York.

Spence, Bill

1980 *Harpooned: The Story of Whaling*. Crescent Books, New York.

Spoehr, Alexander

1988 19th Century Chapter in Hawai'i's Maritime History: Hudson's Bay Company Merchant Shipping, 1829–1859. *Hawaiian Journal of History* 22:70–100.

Stackpole, Edouard A.

1953 *The Sea Hunters: The New England Whalemen during Two Centuries 1635–1835*. Lippincott, Pennsylvania.

1967 *The* Charles W. Morgan*: The Last Wooden Whaleship*. Meredith Press, New York.

1972 *Whales and Destiny: The Rivalry between America, France and Britain for Control*. University of Massachusetts Press, Amherst.

Stanbury, Myra

2010 Slate Log, *Lady Lyttleton*. *Newsletter of the Australasian Institute for Maritime Archaeology* 29(2):15.

2015 Mermaid Atoll Shipwreck: A Mysterious Early 19th-Century Loss. *Australian National Center of Excellence for Maritime Archaeology and the Australian Institute for Maritime Archaeology Special Publication*, Freemantle, Australia.

Staniforth, Mark

1985 The Introduction and Use of Copper Sheathing—A History. *Bulletin of the Australian Institute for Maritime Archaeology* 9(1/2):21–48.

1987 The Casks from the Wreck of the "William Salthouse." *Australian Journal of Historical Archaeology* 5:21–28.

2000 A Future for Australian Maritime Archaeology? *Australian Archaeology* 50(1):90–93.

Starbuck, Alexander

1878 *History of the American Whale Fishery from its Earliest Inception to the Year 1876.* Castle Books, Secaucus, New Jersey.

State Street Trust and Walton Advertising and Printing (pub.)

1915 *Whale Fishery of New England; an Account, with Illustrations and Some Interesting and Amusing Anecdotes, of the Rise and Fall of an Industry Which Has Made New England Famous Throughout the World.* Boston.

Steel, David

1794 *The Elements and Practice of Rigging, Seamanship, and Naval Tactics.* Reprinted 2011, Vol. 4. Cambridge University Press, Cambridge, UK.

Steele, R.

2005 The Industrial Technology of Shore Whaling in New Zealand. Unpublished master's Thesis, Department of Anthropology, University of Otago, Dunedin, New Zealand.

Steffy, J. Richard

1994 *Wooden Ship Building and the Interpretation of Shipwrecks.* Texas A&M Press, College Station.

Stein, Douglas L.

1992 *American Maritime Documents, 1776–1860, Illustrated and Described.* Mystic Seaport Museum, Mystic, Connecticut.

Stevens, Willis

1997 Red Bay. In *Encyclopedia of Underwater and Maritime Archaeology*, edited by James P. Delgado, pp. 336–338. Yale University Press, New Haven, Connecticut.

Stoliar, Abram

2001 Milestones of Spiritual Evolution in Prehistoric Karelia. *Folklore: Electronic Journal of Folklore* 18–19:80–126.

Stone, David L.

1993 *The Wreck Diver's Guide to Sailing Ship Artifacts of the 19th Century.* Underwater Archaeology Society of British Columbia, Vancouver, Canada.

Strother, Eric, Allen Estes, Aimee Arrigoni, James M. Allen, and William Self

2007 *Final Archaeological Resources Report 300 Spear Street Project San Francisco, California.* Report on file, Planning Department, City and County of San Francisco, California.

Sturtevant, G. C.

1955 Scrimshaw. *Nautical Research Journal* 7(1–2):7–10.

Sussman, Lynne

1977 Changes in Pearlware Dinnerware, 1780–1830. *Historical Archaeology* 11(1):105–111.

Symonds, James, and Eleanor C. Casella

2006 Historical Archaeology and Industrialization. In *The Cambridge Companion to Historical Archaeology*, edited by Dan Hicks and Mary C. Beaudry, pp. 143–167. Cambridge University Press, Cambridge, United Kingdom.

Taeaber, Irene B.

1962 Hawaii. *Population Index* 28(2):97–125.

Tanner, Henry S.

n.d. Chart of the Pacific Ocean. G9230 18—.T2 TIL. Library of Congress Geography and Map Division, Washington, DC.

Tate, Merz

1980 *The United States and the Hawaiian Kingdom: A Political History*. Yale University Press, New Haven, Connecticut.

Taylor, Albert P.

1929 Lahaina: The Versailles of Old Hawaii. *Thirty-Seventh Annual Report of the Hawaiian Historical Society for the Year 1928*. Hawaiian Historical Society, Honolulu.

Taylor, Clarice B.

1952 *Tales of Hawaii: The James Robinson Family*. Mark Alexander Robinson Trust, Honolulu.

Taylor, George Rogers

1977 Nantucket Oil Merchants & the American Revolution. *Massachusetts Review* 18(3):581–606.

Taylor, J. Garth

1988 Labrador Inuit Whale Use in the Early Contact Period. *Arctic Anthropology* 25(1):120–130.

Teague, George Allen

1987 The Archaeology of Industry in North America. Unpublished PhD dissertation, Department of Anthropology, University of Arizona, Tucson.

Temperance Advocate and Seamen's Friend [Honolulu, Hawaii]

1843 Whaleship *Parker* Wrecked on Ocean Island. June 27, 1(6):28.

Thomas, Lindsey

2006 *Site Formation Processes: Preliminary Study for Wooden Shipwrecks in the Northwestern Hawaiian Islands*. Report on file, Northwestern Hawaiian Islands Marine National Monument.

Thomson, Basil

1914 Lost Explorers of the Pacific. *Geographical Journal* 44(1):12–29.

Thomson, Lyndall

1997 The Biodegradation of the Wreck *Day Dawn*. *Bulletin of the Australian Institute for Maritime Archaeology* 21:119–124.

Thrower, Norman J. W.

1967 The Art and Science of Navigation in Relation to Geographical Exploration Before 1900. In *The Pacific Basin. A History of Its Geographical Exploration*, edited by H. K. Friis, pp. 18–39. American Geographical Society Special Publication 38. Burlington, Vermont.

Thrum, Thos G.

1915 *Hawaiian Almanac and Annual: The Reference Book of Information and Statistics Relating to the Territory of Hawaii, or Value to Merchants, Tourists and Others*. Published by author, Honolulu, Hawaii.

Thurn, Everard I.

1915 European Influence in the Pacific, 1513–1914. *Geographical Journal* 45(4):301–322.

Tompkins, E. Berkeley

1972 Black Ahab: William T. Shorey, Whaling Master. *California Historical Quarterly* 51(1):75–84.

Tønnessen, Johan N., and Arne O. Johnsen

1982 *The History of Modern Whaling*. University of California Press, Berkeley.

Toonen, Robert J., Kimberly R. Andrews, Iliana B. Baums, Christopher E. Bird, Gregory T. Concepcion, Toby S. Daly-Engel, Jeff A. Eble, Anuschka Faucci, Michelle R. Gaither, and Matthew Iacchei.

2011 Defining Boundaries for Ecosystem-based Management: A Multispecies Case Study of Marine Connectivity across the Hawaiian Archipelago. *Journal of Marine Biology* 2011:1–22.

Tower, Walter S.

1907 *A History of the American Whale Fishery*, Vol. 20. Publications of the University of Pennsylvania, Philadelphia.

Trinder, Barrie

1982 *The Making of the Industrial Landscape*. J. M. Dent & Sons, London.

Tuttle, Michael C., Norine Carroll, Kelly Bumpass, Michael Krivor, and Jason Burns

2010 2008 Diver Investigations of the Suspected Serapis Site, Ambodifototra, Isle Ste Marie, Madagascar. Plymouth, New Hampshire.

US Coast and Geodetic Survey

1919 *Pilot Guides*. US Government Printing Office, Washington, DC.

US Fish and Wildlife Service

2008 *National Oceanic and Atmospheric Administration, and State of Hawai'i Department of Land and Natural Resources*. Report on file, Papahanaumokuakea Marine National Monument, Honolulu, Hawaii.

Vance, Justin W.

2002 The American Civil War and the Kingdom of Hawaii: Islands in the Wake. Unpublished master's thesis, Department of History, Hawaii Pacific University, Honolulu.

Vancouver, George

1798 *A Voyage of Discovery to the North Pacific Ocean, and Round the World*. G. G. and J. Robinson and J. Edwards. London.

Van Tilburg, Hans

2002a *Maritime Cultural Resources Survey: Northwestern Hawaiian Islands*. Northwestern Hawaiian Islands Coral Reef Ecosystem Reserve.

2002b Underwater Archaeology, Hawaiian Style. In *International Handbook of Underwater Archaeology*, edited by Carol V. Ruppé and Janet F. Barstad, pp. 247–266. Kluwer Academic/Plenum Publishers, New York.

2003 *Kure and Midway Atoll Maritime Heritage Survey 2003*. Report on file, Northwestern Hawaiian Islands Coral Reef Ecosystem Reserve.

2005 *Maritime Heritage Resources Assessment 2005 Activity Report*. Report on file, Hawaiian Islands National Wildlife Refuge, US Fish and Wildlife Service, Department of the Interior.

2006 *Maritime Heritage Survey NWHI June 23–July 20, 2006 Post Cruise Project*

Activities Report. Report on file, Northwestern Hawaiian Islands Marine National Monument.

2007 *Activity Report: Maritime Heritage Resources Survey HI-07–07*. Report on file, Northwestern Hawaiian Islands Marine National Monument.

2010 *A Civil War Gunboat in Pacific Waters: Life Aboard the USS* Saginaw. University Press of Florida, Gainesville.

Van Tilburg, Hans, and Kekuena Kikiloi

2007 Papahānaumokuākea Marine National Monument. In *Fathoming Our Past: Historic Contexts of the National Marine Sanctuaries*, edited by Bruce G. Terrell, pp. 56–59. NOAA National Marine Sanctuary Program, Silver Spring, Maryland.

Van Tilburg, Hans, and Joylynn Oliveira

2007 Hawaiian Islands Humpback Whale National Marine Sanctuary. In *Fathoming Our Past: Historic Contexts of the National Marine Sanctuaries*, edited by Bruce G. Terrell, pp. 52–55. NOAA National Marine Sanctuary Program, Silver Spring, Maryland.

Verrill, A. Hyatt

1916 *The Real Story of the Whaler*. D. Appleton, New York.

Veth, Peter

2006 Theoretical Approaches. In *Maritime Archaeology: Australian Approaches*, edited by Mark Staniforth and Michael Nash, pp. 13–26. Springer, New York.

Veth, Peter, and Michael McCarthy

1999 Types of Explanation in Maritime Archaeology: The Case of the SS *Xantho*. *Australian Archaeology* 48(1):12–15.

Vickers, Daniel

1985 Nantucket Whalemen in the Deep-Sea Fishery: The Changing Anatomy of an Early American Labor Force. *Journal of American History* 72(2):277–296.

Vickers, Daniel, and Vince Walsh

2005 *Young Men and the Sea: Yankee Seafarers in the Age of Sail*. Yale University Press, New Haven, Connecticut.

Vosmer, Tom, and Jim Wright

1991 *Lady Lyttlelon*—A Search for Origins. *Bulletin of the Australian Institute for Maritime Archaeology* 15(1):19–30.

Wadell, Peter

1985 The Pump and Pump Well of a 16th-Century Galleon. *International Journal of Nautical Archaeology and Underwater Exploration* 14(3):243–259.

1986 The Disassembly of a 16th-Century Galleon. *International Journal of Nautical Archaeology and Underwater Exploration* 15(2):137–148.

Walker, F. D.

1909 *Log of the* Kaalokai. Hawaiian Gazette, Honolulu.

Ward, R. Gerard

1960 *American Activities in the Central Pacific, 1790–1870: A History, Geography, and Ethnography Pertaining to American Involvement and Americans in the Pacific Taken from Contemporary Newspapers Etc.* Reprinted 1967 by Gregg Press, Ridgewood, New Jersey.

Watson, Norman

2003 *The Dundee Whalers.* Tuckwell Press, East Lothian, United Kingdom.

Wede, Karl

1972 *The Ship's Bell: Its History and Romance.* South Street Seaport Museum, New York.

Weiss, Harry Bischoff, Howard R. Kemble, and Millicent T. Carré

1974 *Whaling in New Jersey.* New Jersey Agricultural Society, Trenton.

Wesley, Daryl Lloyd, Sue O'Connor, and Jack N. Fenner

2016 Re-evaluating the Timing of the Indonesian Trepang Industry in Northwest Arnhem Land: Chronological Investigations at Malara (Anuru Bay A). *Archaeology in Oceania* 51(3):169–195.

Weslowski, J.

1978 *Whaleship Tryworks*—(Charles W. Morgan). Mystic Seaport Record Memorandum, Mystic, Connecticut.

Westerdahl, Christer

1992 The Maritime Cultural Landscape. *International Journal of Nautical Archaeology* 21(1):5–14.

Whalemen's Shipping List

1843 Loss of the Whaleship *Parker.* November 7, 1(35):310–311.

1845 Visit of Brig *Delaware* to Pell's Island. July 15, 3(19):75.

Whaling M. S. Rotch Papers

1820 Construction and Outfitting of *William Rotch* ca. 1820. File 23.26.3. New Bedford Whaling Museum Research Library and Archives, New Bedford, Massachusetts.

Whipple, Addison B. C.

1954 *Yankee Whalers in the South Seas.* Victor Gollancz, London.

1979 *The Whalers.* Time-Life Books, Amsterdam.

Whitecar, William B., Jr.

1864 *Four Years Aboard the Whaleship.* J. B. Lippincott, Philadelphia.

Whitehead, Hal

2003 *Sperm Whales: Social Evolution in the Ocean.* University of Chicago Press, Chicago.

Wilde-Ramsing, Mark U., and Charles R. Ewen

2012 Beyond Reasonable Doubt: A Case for *Queen Anne's Revenge. Historical Archaeology* 46(2):110–133.

Williams, Kristin

1997 Management of Wrecks "In the Way," Discussion of the Move of the *Day Dawn. Bulletin of the Australian Institute for Maritime Archaeology* 21(1/2):125–128.

Wilson Barker, David, and William Allingham

1896 *Navigation: Practical and Theoretical.* C. Griffin, London.

Wood, D.

[1873] Abstracts of Whaling Voyages from the United States 1831–1873. New Bedford Whaling Museum, New Bedford, Massachusetts.

Woodward, C. Vann (ed.)

1968 *The Comparative Approach to American History*. Basic Books, New York.

Woodward, Paul W.

1972 *The Natural History of Kure Atoll, Northwestern Hawaiian Islands*. Atoll Research Bulletin No. 164. Smithsonian Institution, Washington DC.

Works Progress Administration (WPA)

1938 *Whaling Masters*. Old Dartmouth Historical Society Whaling Museum, New Bedford, Massachusetts.

1940 *Ship Registers of New Bedford, Massachusetts*, Vol. 1, *1796–1850*. National Archives Project, Boston.

Wray, Phoebe, and Kenneth R. Martin

1979 Historical Whaling Records from the Western Indian Ocean. *Report of the International Whaling Commission* 5:213–241.

Wylie, Robert Crichton

1845 Notes on the Sandwich, or Hawaiian Islands. *Simmonds's Colonial Magazine* 5(May–Aug):253–268.

Young, Lucien

1890 *Simple Elements of Navigation*. John Wiley & Sons, New York.

Index

Page numbers in italics refer to figures.